P9-DCM-479

Energy and Power

A SCIENTIFIC AMERICAN *Book*

Energy and Power

W. H. FREEMAN AND COMPANY
San Francisco

Copyright © 1971 by Scientific American, Inc.
All rights reserved. No part of this book may be reproduced by any mechanical, photographic, or electronic process, or in the form of a phonographic recording, nor may it be stored in a retrieval system, transmitted, or otherwise copied for public or private use without written permission from the publisher.

The eleven chapters in this book originally appeared as articles in the September 1971 issue of *Scientific American*. Each is available as a separate Offprint from W. H. Freeman and Company, 660 Market Street, San Francisco, California 94104.

Library of Congress Catalogue Card Number:
75-180254

International Standard Book Number:
0-7167-0939-2 (cloth)
0-7167-0938-4 (paper)

Printed in the United States of America

9 8 7 6 5

Contents

Foreword

This book addresses the frightening import of the simple mathematical tautology that is implicit in an exponential curve. The doubling of any quantity—whether population, food production or energy requirements—at regular, say ten-year, intervals means that the magnitude of the last term in the series must exceed the cumulative total of all the terms that precede it. Thus, in the series 1, 2, 4, 8, the 8 exceeds the sum of 1 + 2 + 4.

Since the turn of the century, the people of the United States have been doubling their consumption of electrical energy every decade. Suddenly, the supply of energy is beset with crisis. Brown-outs and black-outs of public utility systems, fuel shortages and increases in fuel prices, local skirmishes over the siting of fossil-fuel and nuclear plants, oil spills from tankers and offshore wells, public anxiety excited by these accidents and by generalized concern for "ecology"—all have taken the magic out of the finger on the light switch, the foot on the accelerator. With the present doubling it can be seen now that energy is not, after all, infinitely and instantly available to man's will.

The energy flux is the common denominator of all natural and human systems. In the absence of life, energy everywhere in the universe courses downhill, in accordance with the second law of thermodynamics, from high-potential sources at temperatures of millions of degrees in the stars and in the infernos of galactic nuclei, down to temperatures near absolute zero in the void of space. Life on earth—and perhaps on planets of other suns—traps a tiny portion of the solar flux in the chemical bonds of large and intricately structured molecules, staying the passage of energy momentarily in its outward and downward flight. In natural communities, an increasing diversity of living forms captures and conserves the solar input; at ecological "climax" both the diversity of life and the capture and conservation of energy reach a maximum. Then the system returns to the sky, from day to day or season to season, as much energy as the sun pours in.

Man first appeared in the natural order as a creature of the climax; his presence as a hunter and food-gatherer left the balance of input and output—photosynthesis and respiration—undisturbed. With the agricultural-urban revolution, beginning about ten thousand years ago, human settlements reduced the high diversity and efficiency of the natural communities in order to store solar energy in plant and animal tissue edible and otherwise usable by man. From day to day and season to season, the inflow and outflow of energy remained in balance, but less of the flux was shunted through life processes.

Industrial civilization upsets the energy balance. To the daily or seasonal return of solar energy to space, it adds the flux of solar energy stored in fossil fuels. Within a time of no more than half a millennium, if industrial civilization continues to burn fossil fuels, this ancient credit to the earth's energy account, built up over geological ages, will have been expended.

Although the nuclear energy production cycles lift the resource horizon, the consumption of energy must ultimately encounter another constraint: the provision not of source but of sink. In acute and local form, this question is already pressed by the heat pollution from electrical power stations on inland streams and coastal waters and by the perturbation of local microclimates by the heat and moisture issuing from such plants, their condensing towers and the communities they serve. An only partly facetious extrapolation of present trends to the year 2100 shows the United States radiating into space as much heat from fossil and nuclear fuels as it receives from the sun.

This book places the energy consumption of civilization in a long time perspective, reaching into the future as well as into the past, and in the broad context of understanding that comes—along with power itself—from objective knowledge. The chapters in this book were first published in the September 1971 issue of *Scientific American*, which was the twenty-second in the series of single-topic issues published annually by the magazine. The editors herewith express appreciation to their colleagues at W. H. Freeman and Company, the book-publishing affiliate of *Scientific American*, for the enterprise that has made the contents of this issue so speedily available in book form.

THE EDITORS*

September, 1971

*BOARD OF EDITORS: Gerard Piel (Publisher), Dennis Flanagan (Editor), Francis Bello (Associate Editor), Philip Morrison (Book Editor), Jonathan B. Piel, John Purcell, James T. Rogers, Armand Schwab, Jr., C. L. Stong, Joseph Wisnovsky

I

Energy and Power

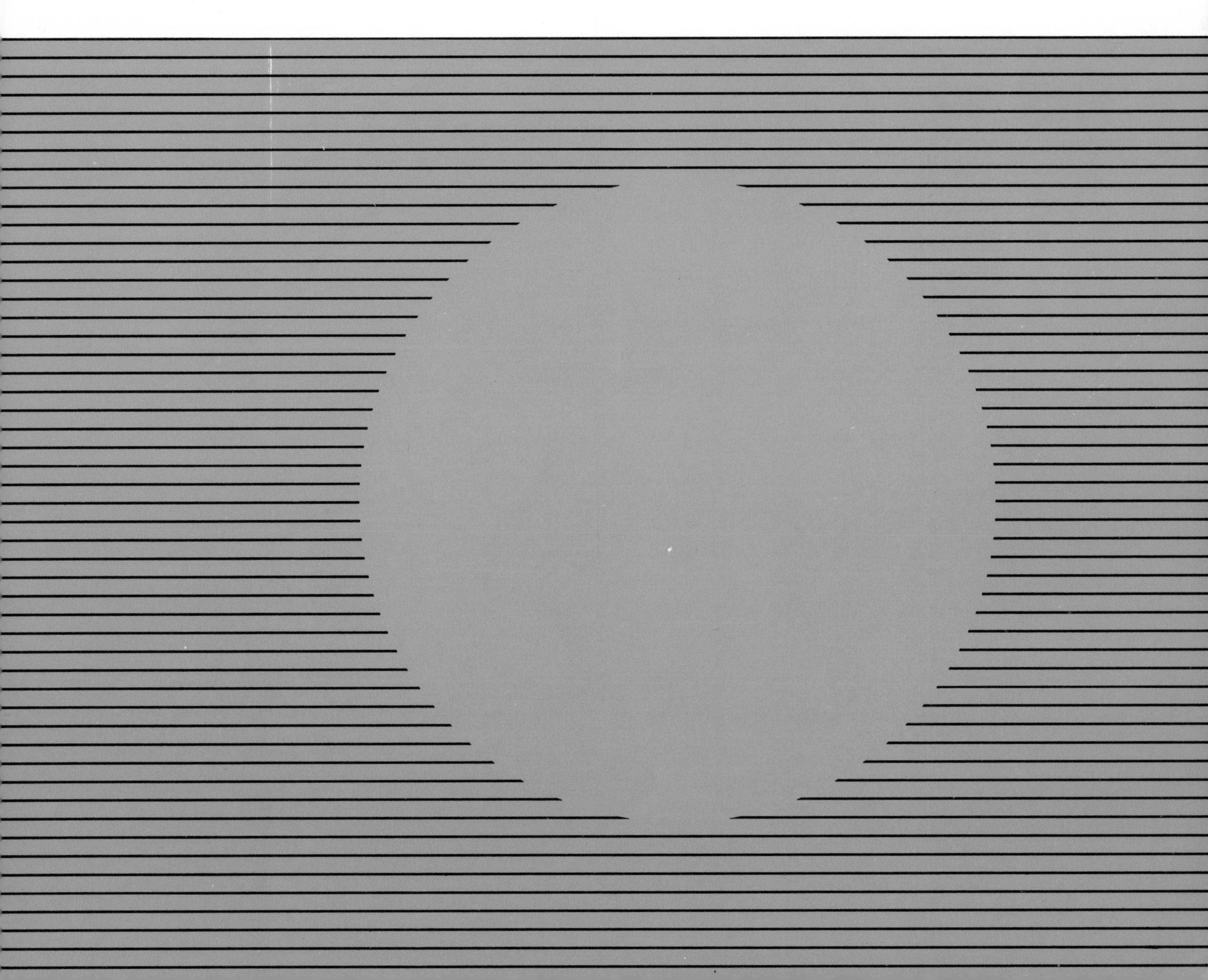

Energy and Power

CHAUNCEY STARR

Man's expanding need for energy creates difficult economic, social and environmental problems. The solutions call for sensible choices of technological alternatives by the market and political process

Between now and 2001, just 30 years away, the U.S. will consume more energy than it has in its entire history. By 2001 the annual U.S. demand for energy in all forms is expected to double, and the annual worldwide demand will probably triple. These projected increases will tax man's ability to discover, extract and refine fuels in the huge volumes necessary, to ship them safely, to find suitable locations for several hundred new electric-power stations in the U.S. (thousands worldwide) and to dispose of effluents and waste products with minimum harm to himself and his environment. When one considers how difficult it is at present to extract coal without jeopardizing lives or scarring the surface of the earth, to ship oil without spillage, to find acceptable sites for power plants and to control the effluents of our present fuel-burning machines, the energy projections for 2001 indicate the need for thorough assessment of the available options and careful planning of our future course. We shall have to examine with both objectivity and humanity the necessity for the projected increase in energy demand, its relation to our quality of life, the practical options technology provides for meeting our needs and the environmental and social consequences of these options.

The artful manipulation of energy has been an essential component of man's ability to survive and to develop socially. Although primitive people and most animals can alter their behavior to adapt to changing environmental restrictions, the reverse ability to substantially alter the environment is uniquely man's. When primitive man learned to use fire to keep himself warm, he took the first big step in the use of an energy resource.

The use of energy has been a key to the supply of food, to physical comfort and to improving the quality of life beyond the rudimentary activities necessary for survival. The utilization of energy depends on two factors: available resources and the technological skill to convert the resources to useful heat and work. Energy resources have always been generally available, and the heating process is ancient. Power devices able to convert energy into useful work have been a recent historical development. The prehistoric domestication of animals represented a multiplication in the power resources available to man, but not by very significant amounts. The big importance of the horse and the ox was that their fuel requirements did not deplete man's own food supply. During this period the power available limited man's ability to irrigate, cultivate and survive.

Water power for irrigation purposes, exploiting natural differences in elevation, was known in very early times. The horizontal waterwheel appeared about the first century B.C. with a power capacity of perhaps .3 kilowatt. By the fourth century the vertical waterwheel had been developed to about two kilowatts of power. These wheels were primarily used for grinding cereals and similar mechanical tasks. By the 16th century the waterwheel was by far the most important prime mover, providing the foundation for the industrialization of western Europe. By the 17th century its power output was reaching significant levels. The famous Versailles waterworks at Marly-la-Machine is said to have had a power of 56 kilowatts. The windmill first appeared in western Europe in the 12th century. It was variously used for grinding grain, for hoisting materials from mines and for pumping water. The windmill had a respectable capacity ranging from several kilowatts to as much as 12 kilowatts. The biggest disadvantage was the intermittent nature of its operation.

BAYWAY REFINERY of the Humble Oil & Refining Company in Linden, N.J., occupies most of the land area in the aerial photograph on the opposite page. Placed on-stream in 1909, when oil supplied less than 6 percent of the nation's energy (it now supplies 43 percent), the refinery has grown with the demand for petroleum products. One of five refineries operated by Humble in the U.S., the Bayway plant refines 200,000 barrels of crude oil per day. Most of it is delivered by tanker and unloaded at docks bordering Arthur Kill, a waterway that separates Staten Island, at the top of the picture, from New Jersey. The multilane highway that cuts across the photograph at an angle is the New Jersey Turnpike.

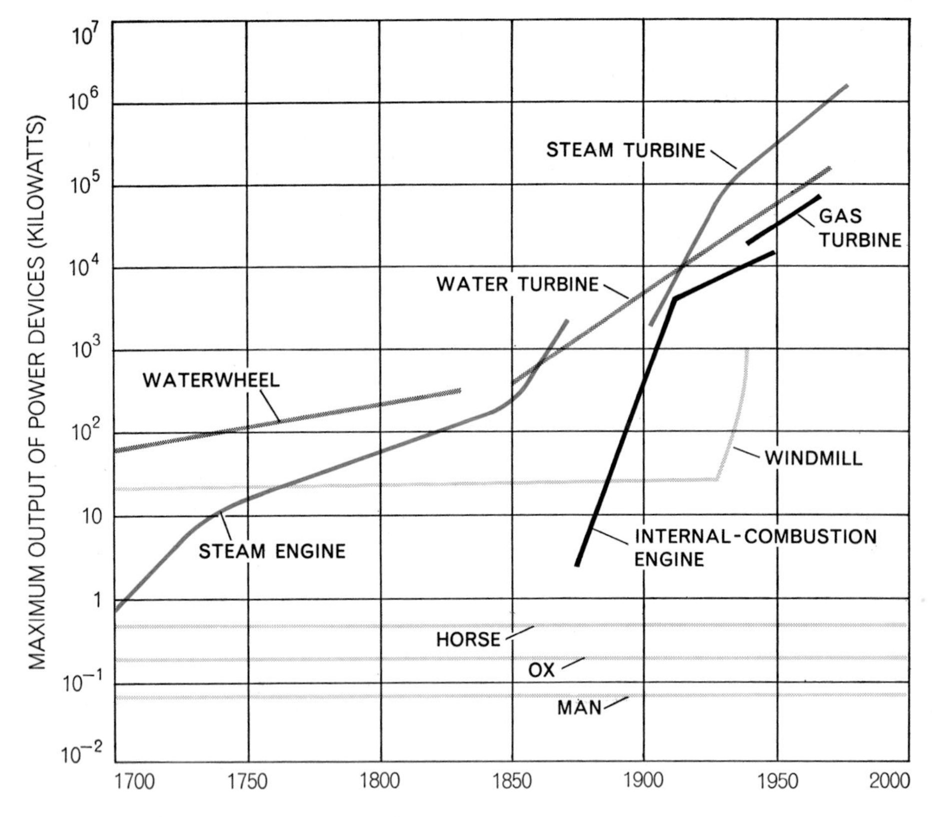

POWER OUTPUT OF BASIC MACHINES has climbed more than five orders of magnitude since the start of the Industrial Revolution (*ca.* 1750). For the steam engine and its successor, the steam turbine, the total improvement has been more than six orders, from less than a kilowatt to more than a million. All are surpassed by the largest liquid-fuel rockets (*not shown*), which for brief periods can deliver more than 16 million kilowatts.

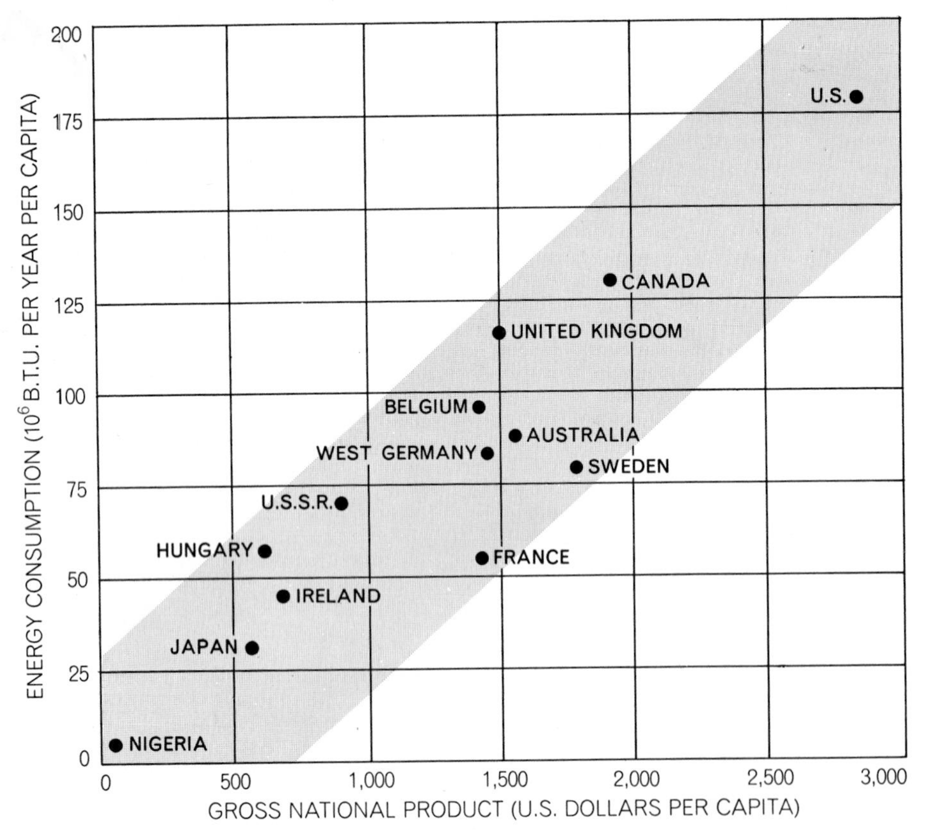

COMMERCIAL ENERGY USE AND GROSS NATIONAL PRODUCT show a reasonably close correlation. A more complete listing of countries is presented in the illustration on page 90 in the article by Earl Cook titled "The Flow of Energy in an Industrial Society."

The development of the steam prime mover is relatively modern compared with the windmill and the waterwheel. As early as the first century after Christ, Hero of Alexandria demonstrated the famous Sphere of Aeolus, a steam reaction turbine on a toy scale. Not until the 17th century was steam used effectively. The steam pump invented by Thomas Savery was a pistonless device using the vacuum of condensing steam to pump water, with a power output of about three-fourths of a kilowatt. Early in the 18th century steam engines using a moving piston were developed as power sources of several kilowatts.

The steam engine was the first mechanical prime mover to provide basic mobility. It was some time, however, before this mobility was used. The early Industrial Revolution was based on the waterwheel and the windmill as prime movers: the location of industrial centers, factories and cities was primarily determined by the availability of those power sources. It was the geographic limitation on the expansion of water power that gave the steam engine an opportunity to continue the growth of manufacturing centers. The first use of the steam engine was as an auxiliary to the waterwheel: to pump water to an elevation sufficient to increase the wheel's power. It was not until the middle of the 19th century that the steam engine became a principal prime mover for the manufacturing industry of the Western world.

The contribution of large power machines to the social development of man became important after 1700 [*see top illustration at left*]. Since 1900 a steadily growing variety of smaller power-conversion devices have been introduced whose chief virtue is mobility. From 1700 on the power output of energy-conversion devices increased by roughly 10,000 times. Most of this growth occurred in the past century, so that it has had its major impact only recently. It is this technological capability that makes our age historically one of accelerated energy utilization. The development of these prime movers required and supported the technology of iron and steel fabrication, and it involved the rise of the railroads. The consequence of these technological innovations has been an exponential increase in energy consumption.

For the millenniums preceding the 17th century the productivity of man was principally determined by his own labor and by that of domestic animals.

The growth in the world population and the manifestations of greater average affluence all appear to show significant increases in parallel with the growth in energy use. Simultaneously one witnessed rapid developments in learning, in the arts and in technologies of all kinds. Although one must be cautious when dealing with pluralistic and interacting relations, a strong case can be made for the hypothesis that the productive utilization of energy has played a primary role in shaping the science and culture of the past three and a half centuries. This hypothesis is supported by the linear relation one finds today between the per capita consumption of energy for heat, light and work and the per capita gross national product of various nations [*see bottom illustration on opposite page*].

As an example of the effect of power machinery on the productivity of man, the agricultural experience of the U.S. following World War I is much to the point [*see illustrations on page 7*]. The story is told in the following quotation from the *1960 U.S. Yearbook of Agriculture:*

"Horse and mule numbers at that time [1918] were the highest in our history—more than 25 million—but the rate of technological progress had slowed down. The availability of good new land had dwindled to insignificance. One-fourth of the harvested crop acre-

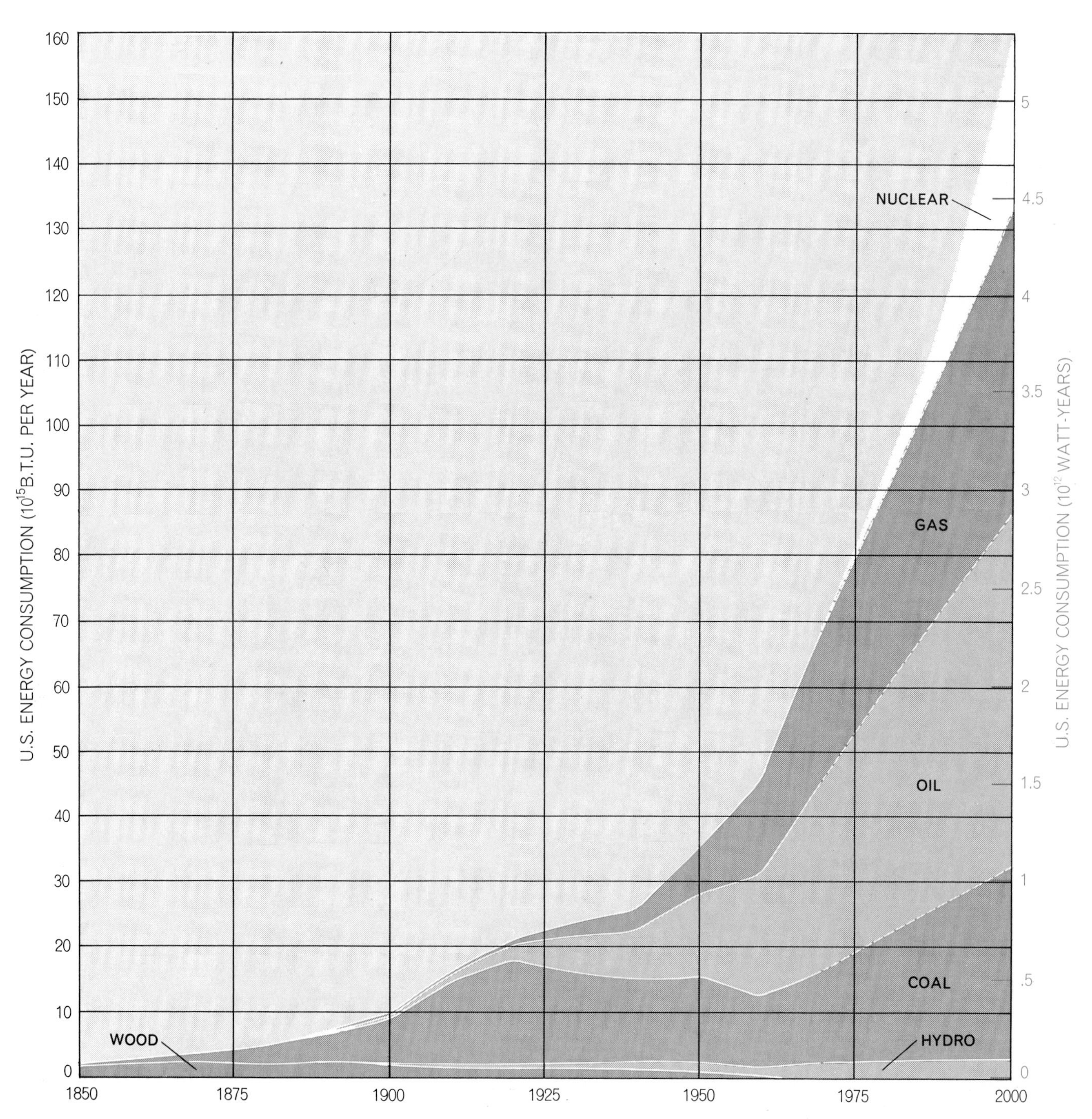

U.S. ENERGY CONSUMPTION has been multiplied some 30 times since 1850, when wood supplied more than 90 percent of all the energy units. By 1900 coal had become the dominant fuel, accounting for more than 70 percent of the total. Fifty years later coal's share had dropped to 36.5 percent and the contribution from oil and natural gas had climbed to 55.5 percent. Last year coal accounted for 20.1 percent of all energy consumed, oil and gas 75.8 percent, hydropower 3.8 percent and nuclear energy .3 percent. Energy-consumption figures are from the U.S. Bureau of Mines; projections conform to those given in the illustration on page 10.

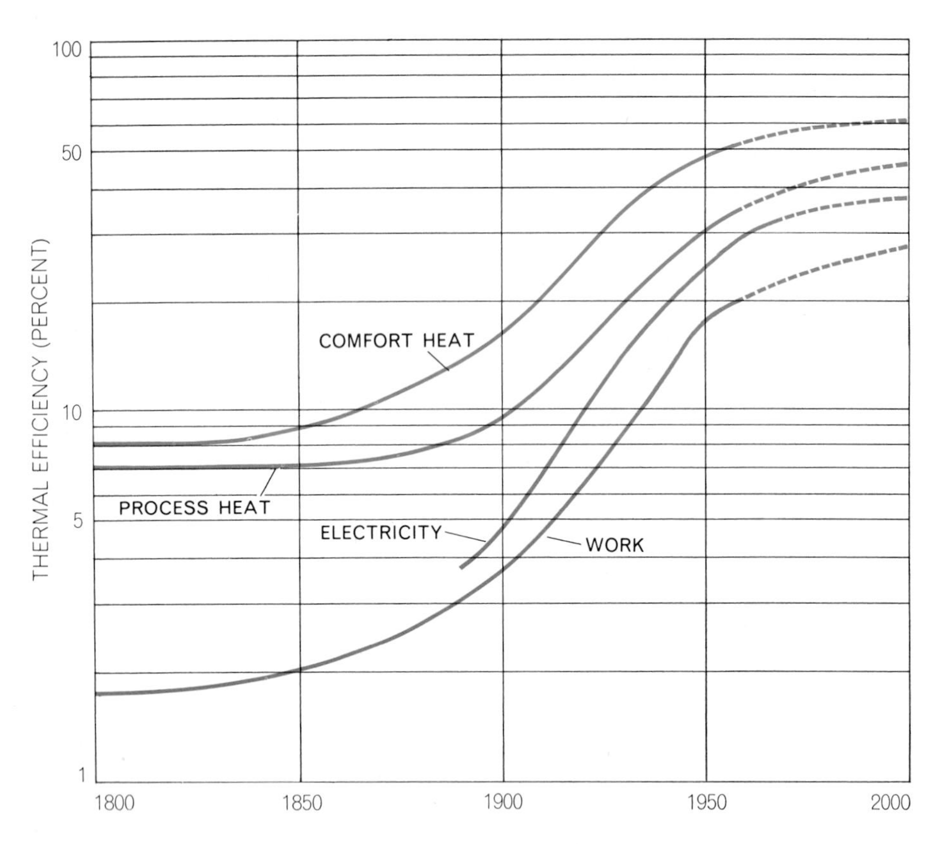

EFFICIENCY OF ENERGY CONVERTERS rose steeply from 1850 to 1950. From here on improvements will be much harder to win, partly because of thermodynamic limitations. A simple unweighted average of efficiencies in four major categories of energy use gives a value of about 8 percent in 1900, 30 percent in 1950 and a projected 45 percent in A.D. 2000.

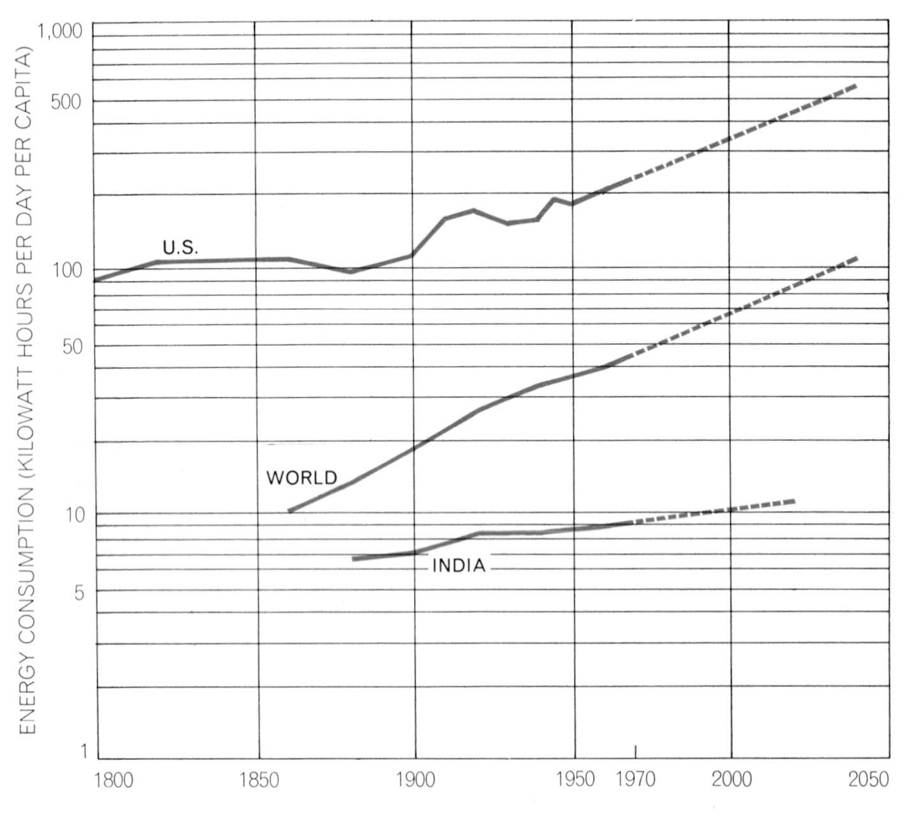

GROWTH IN ENERGY DEMAND in the U.S. is at the annual rate of about 1 percent per capita. For the world as a whole per capita consumption is growing about a third faster. Even so, the world supply of energy per capita in A.D. 2000 will be less than a fourth of the projected U.S. figure. In India the rate of increase is only about a third of the U.S. rate.

age was being used to produce feed for power animals.

"If methods had not been changed, many more horses, more men to work them, and much more land to grow feed for them would be required for today's net agricultural output. The American economy of the 1960's could not be supported by an animal-powered agriculture on our essentially fixed—in fact, slowly shrinking—land base. National progress on all fronts would have been retarded seriously had not agriculture received new forms of power and sources of energy not restricted by biological limitations.

"With the adoption of mechanical forms of power in engines, tractors, and electric motors and development of more and more types of adapted equipment to use that power, American agriculture entered a new era of sharply rising productivity."

The introduction of new hybrid grains, the use of fertilizer and pesticides, along with extensive irrigation, all contributed to the increased productivity per unit of labor. Irrigation systems and the manufacture and transportation of chemical fertilizers on a large scale all require substantial use of energy, as Earl Cook points out in this book ("The Flow of Energy in an Industrial Society," page 83).

It is evident that the present rate of world population growth cannot be sustained indefinitely; sooner or later environmental restrictions will cause the death rate to increase substantially, and the least developed countries will be the first to suffer. The long-term alternative for the world is a controlled birthrate. Nevertheless, for some decades to come social trends will cause an inevitable increase in world population. In order to meet not only the food requirements but also a minimally reasonable quality of life, the contributions that can be made by the use of energy in various forms are essential. The issue therefore is *not* whether energy production for the world should be increased. It is rather how to increase it effectively with minimum deleterious side effects.

Because the great increase in energy consumption in the past century has taken place chiefly in the advanced countries, it is instructive to examine the trends in the U.S. The annual consumption of all forms of energy in the U.S. has increased seventeenfold in the past century, with a corresponding population increase of a little more than fivefold. During this period, in which our

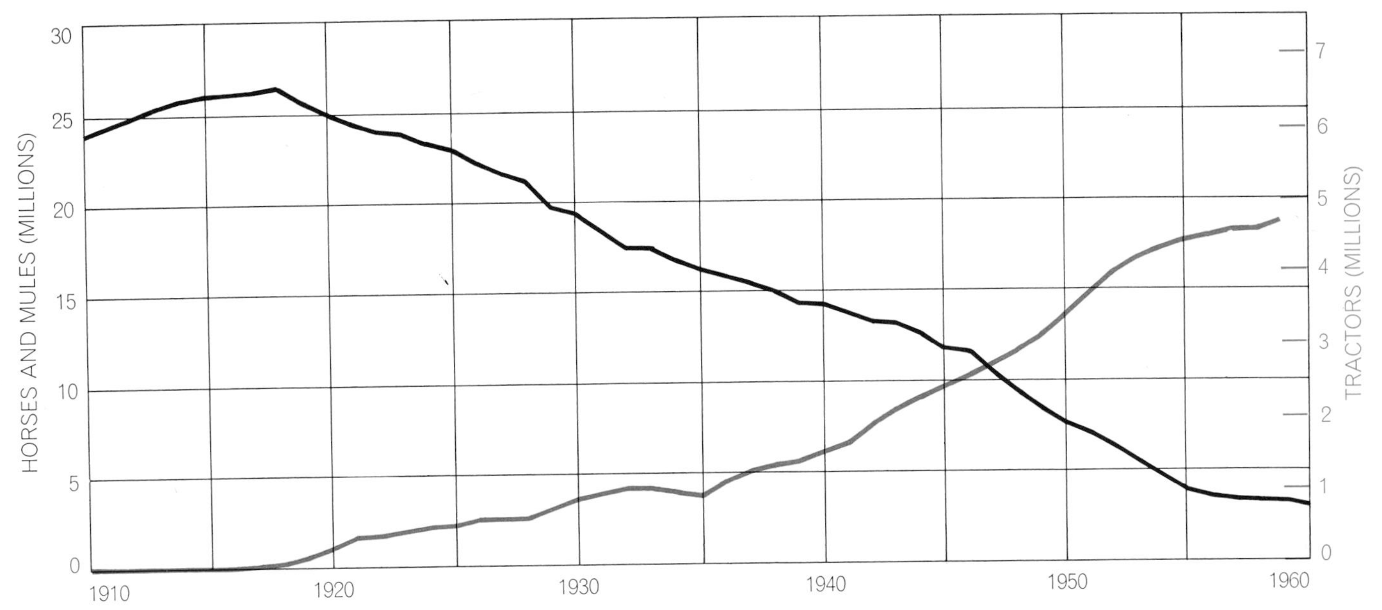

MACHINES REPLACED ANIMALS at a rapid rate on U.S. farms between 1920 and 1960. In the same period farm output more than doubled. In 1920 a fourth of U.S. farm acreage was planted in crops required to feed the nation's 25 million farm horses and mules.

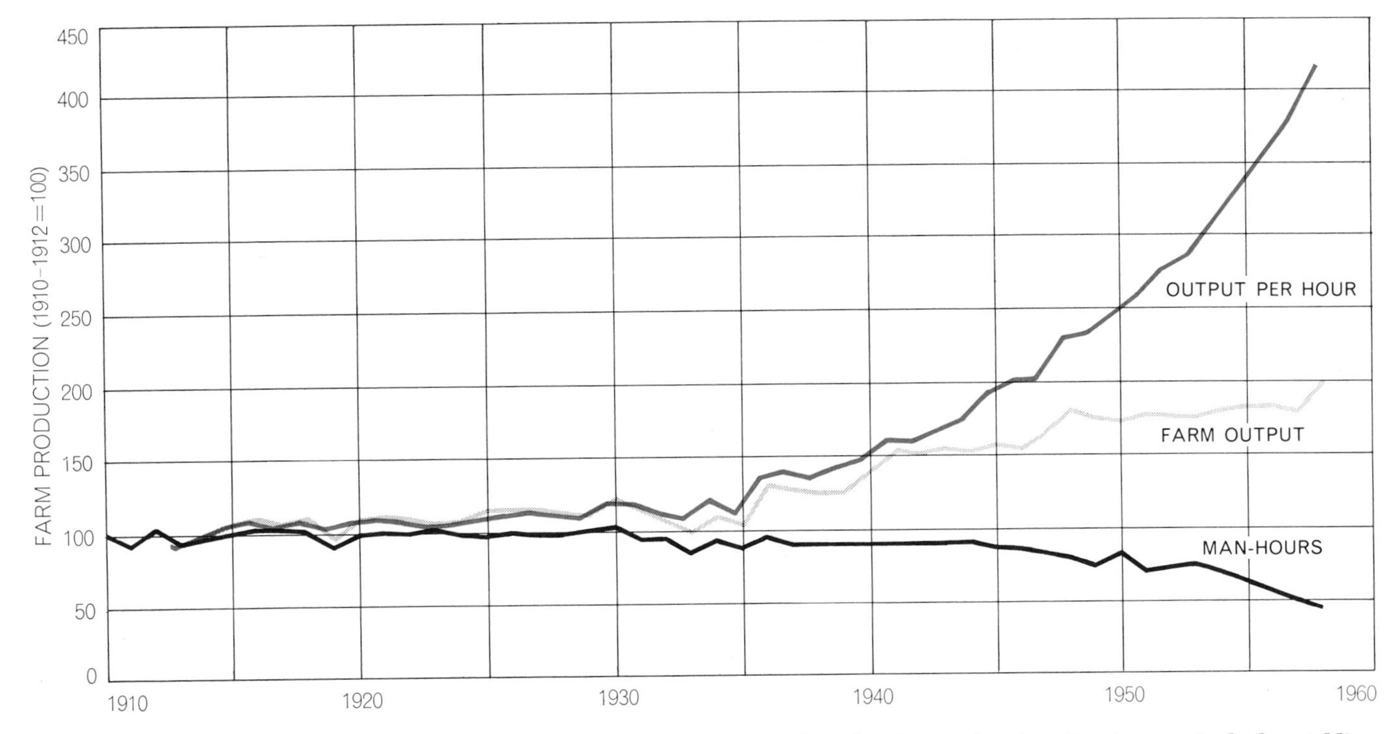

FARM OUTPUT PER MAN-HOUR approximately quadrupled between 1910 and 1958. The improvement was due not only to the internal-combustion engine but also in part to higher-yielding crops, extensive irrigation, fertilizers, herbicides and insecticides.

per capita energy use has slightly more than doubled, fuel sources have shifted steadily [*see illustration on page* 5]. Fuel wood was the dominant energy source in 1850; by 1910 coal accounted for about 75 percent of the total energy consumption and fuel wood had declined to some 10 percent. In the 50 years between 1910 and 1960 coal lost its leading position to natural gas and oil. Today nuclear power is emerging as a national energy source.

Thus roughly 50 years seem to be needed for the energy economy to shift substantially to a new fuel. This is determined primarily by the operating lifetime of power machinery and secondarily by the long lead time for redirecting available manufacturing and supply capability. For example, the steadily increasing demand for electric power requires construction of new power stations at a rate that exceeds the facilities of the infant nuclear-power industry; as a result fossil-fuel-burning plants must be built for many decades to come in order to meet the nation's needs. With an expected operating life of 30 years for such plants, it is evident that fossil-fuel plants will be playing a role even half a century after the change to nuclear power was initiated.

A century ago our energy resources were primarily applied to the production of heat for physical comfort. Less than a quarter of the heat was utilized for metallurgical processes and industrial activities. Today more than half of all energy consumed in the U.S. goes into useful work. Paralleling this shift in the way energy is used there has been a steady improvement in the efficiency with which energy is converted to useful forms [*see top illustration on page 6*]. There is no theoretical limit to the efficiency of energy use for heating. The theoretical limit on the conversion of heat to work is the Carnot thermodynamic efficiency as is explained in this book by Claude M. Summers ("The Conversion of Energy," page 95). The best of our power plants now operate at a thermal efficiency of 40 percent, a figure that may reach 50 percent by the year 2000. Other thermal prime movers are not as efficient. The internal-combustion engine thermal efficiency ranges from 10 to 25 percent, depending on how the engine is used. Because of the impact of efficiency on the economics of use, the motivation for improving efficiency will persist.

At present the U.S. consumes about 35 percent of the world's energy. By the year 2000 the U.S. share will probably drop to around 25 percent, due chiefly to the relative population increase of the rest of the world. The per capita increase in energy consumption in the U.S. is now about 1 percent per year [*see bottom illustration on page 6*]. Starting from a much lower base, the average per capita energy consumption throughout the world is increasing at a rate of 1.3 percent per year. It is evident that it may be another century before the world average even approaches the current U.S. level. At that time the energy gap between the U.S. and the underdeveloped world will still be large. With unaltered trends it would take 300 years to close the gap. By 2000 the world's average per capita energy consumption will have moved only from the present one-fifth of the U.S. average to about one-third of the present U.S. average. Of grave concern is the nearly static and very low per capita energy consumption of areas such as India, a country whose population growth largely negates its increased total production of energy. If the underdeveloped parts of the world were conceivably able to reach by the year 2000 the standard of living of Americans today, the worldwide level of energy consumption would be roughly 10 times the present figure. Even though this is a highly unrealistic target for 30 years hence, one must assume that world energy consumption will move in that direction as rapidly as political, economic and technical factors will allow. The problems implied by this prospect are awesome.

One can better appreciate the energy problem the world faces if one simply compares the cumulative energy demand to the year 2000—when the annual rate of energy consumption will be only three times the present rate—with estimates of the economically recoverable fossil fuels [*see illustrations on opposite page*]. The estimated fossil-fuel reserves are greater than the estimated cumulative demand by only a factor of two. If the only energy resource were fossil fuel, the prospect would be bleak indeed. The outlook is completely altered, however, if one includes the energy available from nuclear power.

There is no question that nuclear power is a saving technical development for the energy prospects for mankind. Promising but as yet technically unsolved is the development of a continuous supply of energy from solar sources. The enormous magnitude of the solar radiation that reaches the land surfaces

NUCLEAR POWER PLANT being built by the Duke Power Company near Clemson, S.C., has three 886,300-kilowatt units in various stages of completion. Unit No. 1 (*right*) is ready to be loaded with fuel; it is expected that it will be supplying power early next year. The three nuclear steam-supply systems were designed and are being manufactured by the Babcock & Wilcox Company. There are now 22 nuclear power plants with a combined capacity of 9,132 megawatts operating in the U.S. Another 99 plants with a capacity of 90,000 megawatts are under construction or on order. By A.D. 2000 nuclear fuels may be supplying half of the nation's electricity.

of the earth is so much greater than any of the foreseeable needs that it represents an inviting technical target. Unfortunately there appears to be no economically feasible concept yet available for substantially tapping that continuous supply of energy. This somewhat pessimistic estimate of today's ability to use solar radiation should not discourage a technological effort to harness it more effectively. If only a few percent of the land area of the U.S. could be used to absorb solar radiation effectively (at, say, a little better than 10 percent efficiency), we would meet most of our energy needs in the year 2000. Even a partial achievement of this goal could make a tremendous contribution. The land area required for the commercially significant collection of solar radiation is so large, however, that a high capital investment must be anticipated. This, coupled with the cost of the necessary energy-conversion systems and storage facilities, makes solar power economically uninteresting today. Nevertheless, the direct conversion of solar energy is the only significant long-range alternative to nuclear power.

The possibility of obtaining power from thermonuclear fusion has not been included in the listing of energy resources on this page because of the great uncertainty about its feasibility. The term "thermonuclear fusion," the process of the hydrogen bomb, describes the interaction of very light atomic nuclei to create highly energetic new nuclei, particles and radiation. Control of the fusion process involves many scientific phenomena that are not yet understood, and its engineering feasibility has not yet been seriously studied. Depending on the process used, controlled fusion might open up not only an important added energy resource but also a virtually unlimited one. The fusion process remains a possibility with a highly uncertain outcome.

The special environmental problems associated with generating electricity have drawn much attention, but the production of electricity is not the major environmental problem we face [*see illustration on next page*]. Of all the energy needs projected for the year 2000, nonelectric uses represent about two-thirds. These uses cover such major categories as transportation, space heating and industrial processes. The largest energy user at that time will be the manufacturing industry, with transportation using about half as much. These projections are based on extrapolations of present trends. One can speculate, however, on major changes in life style or technology that could substantially alter these projections [*see illustration on page 11*]. These hypothetical shifts include all-electric homes, complete air conditioning, more use of electricity in commercial buildings, the electric automobile, the use of electricity in industrial processes, possible large-scale desalination of seawater and, finally, shifting all electricity production to nuclear plants. Such substantial changes could reduce the estimated fossil-fuel requirements in the year 2000 by more than 40 percent, with the greatest component being the shift from fossil to nuclear fuels in generating electricity. Even with such drastic shifts, the total fuel consumed for electricity would still represent no more than 60 percent of the national energy requirement, with the remaining 40 percent still dependent on fossil fuel.

It is clear that if in the year 2000 the U.S. were solely dependent on fossil

DEPLETABLE SUPPLY (10^{12} WATT-YEARS)	WORLD	U.S.
COAL	670 — 1,000	160 — 230
PETROLEUM	100 — 200	20 — 35
GAS	70 — 170	20 — 35
SUBTOTAL	840 — 1,370	200 — 300
NUCLEAR (ORDINARY REACTOR)	~3,000	~300
NUCLEAR (BREEDER REACTOR)	~300,000	~30,000
CUMULATIVE DEMAND 1960 TO YEAR 2000 (10^{12} WATT-YEARS)	350 — 700	100 — 140

ECONOMICALLY RECOVERABLE FUEL SUPPLY is an estimate of the quantities available at no more than twice present costs. U.S. reserves of all fossil fuels are slightly less than a fourth of the world total and its reserves of nuclear fuels are only a tenth of the world total. Fossil-fuel reserves are barely equivalent to twice the cumulative demand for energy between 1960 and 2000. Even nuclear fuel is none too plentiful if one were to use only the ordinary light-water reactors. By employing breeder reactors, however, the nuclear supply can be amplified roughly a hundredfold. (10^{12} watt-years equals 29.9×10^{15} B.t.u.)

CONTINUOUS SUPPLY (10^{12} WATTS)	WORLD		U.S.	
	MAXIMUM	POSSIBLE BY 2000	MAXIMUM	POSSIBLE BY 2000
SOLAR RADIATION	28,000		1,600	
FUEL WOOD	3	1.3	.1	.05
FARM WASTE	2	.6	.2	.00
PHOTOSYNTHESIS FUEL	8	.01	.5	.001
HYDROPOWER	3	1.	.3	.1
WIND POWER	.1	.01	.01	.001
DIRECT CONVERSION	?	.01	?	.001
SPACE HEATING	.6	.006	.01	.001
NONSOLAR				
TIDAL	1.	.06	.1	.06
GEOTHERMAL	.06	.006	.01	.006
TOTAL	18+	3	1.2	.2
ANNUAL DEMAND YEAR 2000 (10^{12} WATTS)	~15		~5–6	

CONTINUOUS, OR RENEWABLE, ENERGY SUPPLY can be divided into two categories: solar and nonsolar. Two sets of estimates are again presented, one for the world and one for the U.S. alone. The figure for total solar radiation includes only the fraction (about 30 percent) falling on land areas. If an efficient solar cell existed to convert sunlight directly to electric power, one could think of utilizing solar energy on a large scale. The sunlight that falls on a few percent of the land area of the U.S. would satisfy most of the energy needs of the country in the year 2000 if converted to electricity at an efficiency of 12 percent.

fuels, the costs of energy would have to increase substantially [*see illustration on page 15*]. The availability of nuclear power will allow these costs to be kept reasonably low. A major reduction in cost will be achieved when the breeder reactor is successfully developed. In a breeder reactor excess neutrons from the fission of uranium 235 are used to convert nonfissionable uranium 238 and thorium 232 into the fissionable isotopes plutonium 239 and uranium 233 respectively. The breeder reactor should make it possible for nuclear fission to supply the world's energy needs for the next millennium. (If the fusion process is ever successful, its cost for electricity would be similar to that of the breeder.) The U.S. Government has recently announced that intensive development of the fast breeder reactor is now national policy. With multimegawatt fast breeders now being constructed in the U.S.S.R. and in western Europe, there appears to be little doubt about their engineering feasibility. The problems now are those of detailed engineering and performance economics.

In the past century the perceived social benefits from the uses of energy

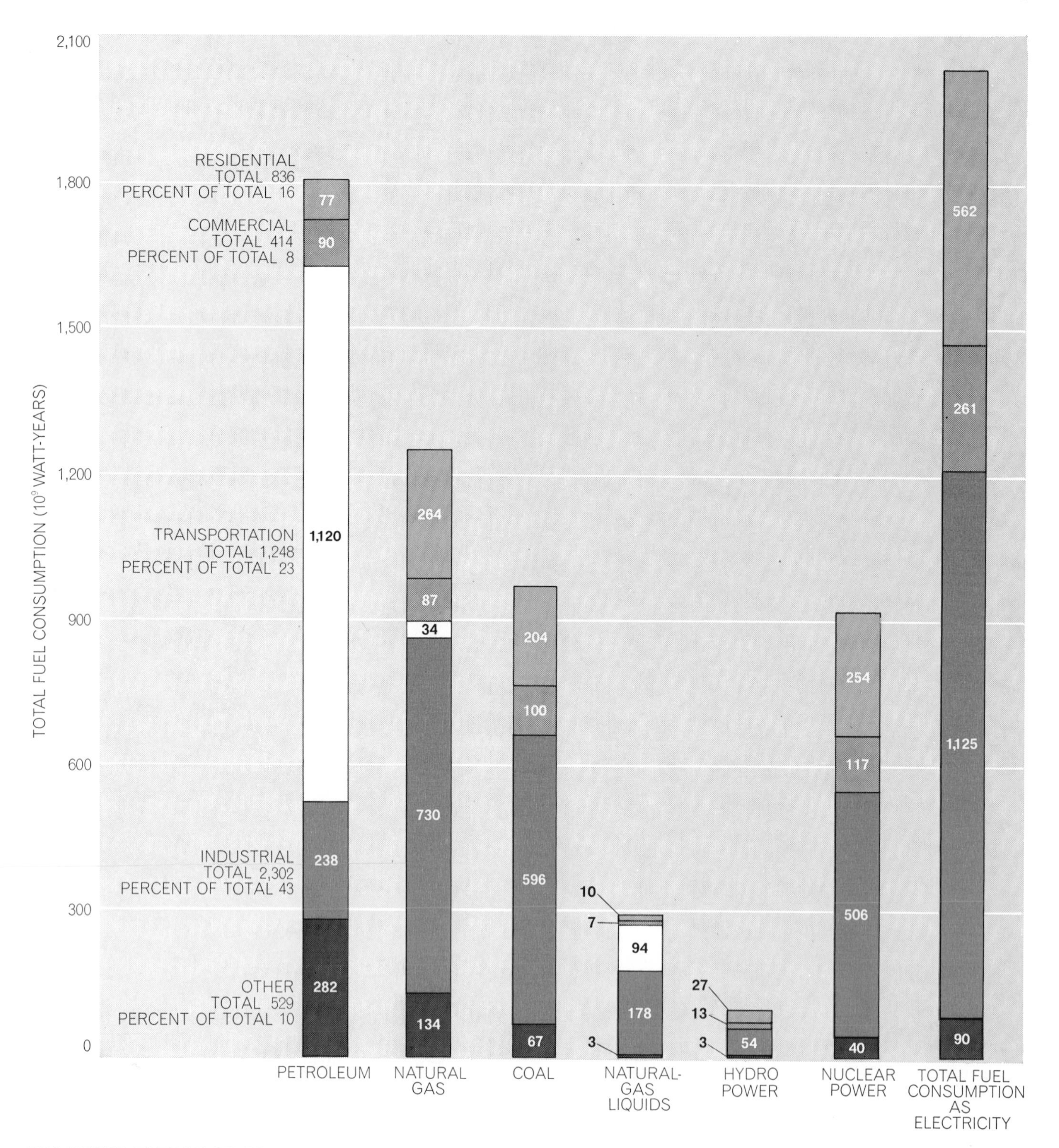

PROJECTED ENERGY SUPPLY AND USE in the U.S. in the year 2000 shows nuclear power contributing almost as much as coal to the total energy supply but both running well behind oil and natural gas. In the year 2000 the generation of electricity may consume 38 percent of the total energy input compared with 25 percent today. These estimates are a projection of present trends. One can imagine, however, a major effort to substitute nuclear for fossil fuels, with the results depicted on the opposite page.

overrode any constraint that might be set by its environmental impact. As the U.S. has grown in both population and affluence, the amount and concentration of our energy use has begun to make the deterioration of the environment serious enough to be of national concern. It is only recently that priority has been given to the technology of pollution abatement; there is little doubt that eventually control of the environmental side effects of energy utilization will be brought to socially acceptable levels. Pollution is man-made and is man-controllable. Pollution control, however, is itself a new growth industry and will create an additional energy demand. Pollution-control techniques use chemical-plant processes, all of which consume energy. For example, the proposed effluent-treatment methods for reducing pollutants from automobiles result in increased fuel consumption.

In considering the harmful effects of energy utilization it is well to distinguish between those that are short-term and geographically concentrated and those that operate over the long term, often with worldwide consequences. Of the latter there are only a few. The combustion of fossil fuels, no matter how efficiently done, must always produce carbon dioxide. Its concentration in the atmosphere has increased from some 290 parts per million to 320 within the past century and may increase to 375 or 400 parts per million by the year 2000. The mechanism for the removal of carbon dioxide is only partly understood; it is eventually absorbed by the ocean, converted into minerals or incorporated by plants in their growth. Thus the carbon dioxide ultimately but slowly returns to the biosphere in some nonpolluting form. Its effects while it resides in the atmosphere are not now predictable, although theoretically the increased carbon dioxide should cause a "greenhouse effect" by reducing the infrared heat loss from the earth and perhaps raising the mean global temperature one degree Celsius by the year 2000.

In parallel with the increase in carbon dioxide in the atmosphere there has also been a rise in suspended particulate contamination. Fine particles are released into the air not only by combustion but also by volcanic eruptions. The increased turbidity reduces solar radiation to the earth's surface. So far the observed temperature trends are not meaningful and the subject is not well understood. The meteorological data available indicate that neither the added carbon dioxide nor the particulates are a serious problem yet. In any case we have at least several decades for determining the carbon dioxide pathways in our biosphere. If the carbon dioxide additions to the atmosphere were determined to be harmful, there is an ultimate but costly technological solution: we could use nuclear electric power to manufacture hydrogen by the electrolysis of water. Hydrogen would make an ideal fuel because its combustion yields water as an end product.

Other pollutants that arise from the burning of fossil fuels are in a somewhat different category. They tend to concentrate in the region where they are generated and have a relatively short life. They all eventually disappear from the atmosphere through photochemical reactions or meteorological processes such as rain. The problems they create in urban areas because of their high concentration are those associated with either material damage, aesthetics, physical discomfort or public health. If one is willing to pay the cost, one can reduce the quantities of various harmful byproducts by changing combustion processes or by instituting effluent controls.

The end product of nuclear fission is an assortment of radioactive isotopes that have a wide range of lifetimes extending up to thousands of years. Although the total radioactivity decreases with time, there is no question that these radioactive substances must be carefully contained, controlled and stored. Fortunately the physical amounts involved are extremely small in bulk: about 10 cubic feet per year from a 1,000-megawatt fast-breeder power plant. The problem is one of extracting these substances during the chemical processes used for reconstituting the nuclear fuels, and then containing and storing them in

	FOSSIL-FUEL REDUCTION (10^9 WATT-YEARS)	INCREASED ELECTRICITY PRODUCTION (10^9 WATT-YEARS)	INCREASED NUCLEAR ELECTRICITY PRODUCTION (10^9 WATT-YEARS)
ALL-ELECTRIC HOMES	230	99	99
100 PERCENT AIR CONDITIONING	NO CHANGE	14	14
INCREASED USE OF ELECTRICITY IN THE COMMERCIAL SECTOR	90	53	52
ELECTRIC AUTOMOBILES	340	92	91
REPLACE 1/3 OF INDUSTRIAL CONSUMPTION OF GAS	200	228	115
POTENTIAL NEED FOR DESALINATION IN THE WESTERN U.S.	NO CHANGE	114	115
ALL ELECTRICITY PRODUCTION SHIFTED TO NUCLEAR	1,040	NO CHANGE	515
SUBTOTALS	(−) 1,900	(+) 600	(+) 1,000
REFERENCE PROJECTION (10^9 WATT-YEARS)	4,315	1,030	515
PERCENT CHANGE	−44	+58	+194

MAJOR EFFORT TO REDUCE FOSSIL-FUEL USE by the year 2000 might conceivably eliminate 1,900 $\times 10^9$ watt-years from the demand of 4,315 $\times 10^9$ watt-years projected in the bar chart on the opposite page. This would amount to a reduction of 44 percent. The next column shows the amount of electrical energy needed to replace fossil fuel in each of the six categories of energy use listed at the left. The total increase in electric demand comes to 600 $\times 10^9$ watt-years, or 58 percent. The reference projection of 1,030 $\times 10^9$ watt-years assumes the conversion of 2,038 $\times 10^9$ watt-years of fuel at a thermal efficiency of 51 percent. The reference projection also assumes that half of the electric-power production, or 515 $\times 10^9$ watt-years, will be nuclear in the year 2000. The last column shows the increase in nuclear power required if all electricity were to be obtained from nuclear fuels.

a safe manner. Because of the small volume of material produced in the annual operation of a nuclear power station even elaborate handling procedures contribute only a small part to the cost of nuclear power. Although the total amount of radioactive waste today is relatively small, the amounts will be large 30 years hence. Pilot programs are needed now to develop safe handling for these future wastes.

All energy use ends up as unrecoverable waste heat. The final heat sink for the earth is radiation to space. The worldwide man-made thermal load, however, is so small compared with the solar heat load as to be insignificant on a global scale. In the year 2000 the worldwide use of energy will still be much less than a thousandth of the sun's heat input. Nevertheless, one can expect that the concentrated generation and consumption of energy in densely populated areas will be capable of affecting both the local climate and ecological systems. Since rationing of energy does not seem feasible, the only practical solution may be to limit the population density of our major cities.

While recognizing the troposphere as the ultimate heat sink, we have a number of options for influencing the flow of heat from the point where it is released to ultimate radiation into space. Of great public importance is the management of the large quantities of waste heat produced in the generation of electricity. It has been customary to locate electric-power stations on large bodies of water, rivers, lakes or oceans, for the purpose of using the available cooling water to reduce the minimum temperature of the Carnot cycle involved in the generation of power. Because of the recent growth in electric-power generation many of the inland bodies of water are approaching a natural limitation in their ability to absorb waste heat. The most severe of such limitations is the ecological effect on marine life; the maximum temperature that can be tolerated by marine animals is not high.

One way to avoid heating inland bodies of water is to use the waste heat to evaporate a relatively small volume of water rather than to raise a large volume by only a few degrees. Evaporation is carried out by means of a "wet" cooling tower, which is now rather widely used by electric-power stations, particularly in Britain. This approach presents two problems. If the water is drawn from a small river or a small lake, the amount evaporated can reduce the amount available for other purposes. The second problem arises from the considerable amount of water vapor added to the atmosphere, which produces a sharp increase in the local humidity. In some regions of the country, valleys in particular, this would produce heavy fog

CONTROL	INDIVIDUAL SELECTION	SOCIETAL SELECTION	ECONOMIC FEASIBILITY	TECHNICAL FEASIBILITY
IMPLEMENTATION TIME (YEARS)	1	10	10 – 100	
COSTS INVOLVED (DOLLARS)	$10^2 - 10^4$	$10^6 - 10^8$	$10^9 - 10^{11}$	
	OPTIONAL USES	**DEVICE UTILIZATION**	**SPECULATIVE RESOURCES**	
	COMFORT (HEATING, AIR CONDITIONING)	CENTRAL STATION V. LOCAL POWER PLANT	SOLAR POWER	
	ENTERTAINMENT	TYPE OF CONVERSION METHOD	FUSION	
	COMMUNICATION	DISTRIBUTION ALTERNATIVES	BIOLOGICAL PHOTOSYNTHESIS	
	HOME	**RESOURCE DEVELOPMENT**	FUEL CELLS, MHD, DIRECT CONVERSION	
	TRANSPORTATION	COAL	**ALTERNATIVE FUELS**	
	LABOR AID	OIL AND NATURAL GAS	ALCOHOL	
	CRITERIA	NUCLEAR	LIQUID HYDROGEN	
	RELATIVE COSTS	SHALE OIL	AMMONIA	
	PERSONAL SAFETY	COAL GASIFICATION	**ENVIRONMENTAL EFFECTS**	
	QUALITY OF LIFE	FUSION	RECYCLE WASTES	
	INTANGIBLE AND SUBJECTIVE BIASES	SOLAR	WASTE STORAGE (RADIOACTIVE)	
		SITING CHOICES	UNDERGROUND DISTRIBUTION	
		AT ORIGIN OF FUEL	SAFETY	
		CLOSE TO USER		
		CONSIDER AESTHETICS		
		LAND-USE ALTERNATIVES		
		WASTE DISPOSAL		
		ENVIRONMENTAL DETERIORATION		
		REGULATION AND CONTROL		
		LEGISLATION		
		REGULATIONS		
		STANDARDS		

CONTROLLING FACTORS that enter into long-range energy planning are listed in this table. Some factors, such as those listed under individual and societal selection, can operate in a relatively short time. Other factors, such as those listed under economic

during much of the year, with considerable discomfort to the local population.

The next choice in heat disposal would be direct dissipation to the air from a closed-cycle heat exchanger in the form of "dry" cooling towers. This is the same technique used in an automobile radiator for cooling the engine. Although dry cooling towers obviate the need for a water supply altogether, they not only require a higher capital investment but also decrease the thermodynamic efficiency of the power station because ambient air temperatures are generally much higher than the temperatures that can be reached with a water-cooling system. Nevertheless, for inland power stations environmental considerations may force a steady increase in the use of dry cooling towers.

In many respects the most suitable location for electric-power stations is on or near the ocean. The ocean represents a heat sink of such magnitude as to be on the average unaffected by the waste heat man can introduce for the foreseeable future. There are also many areas of the ocean where local increases in temperature could even be beneficial to marine life. For this reason the location of power stations on the shores of large oceans may become increasingly popular throughout the world.

It is evident that the issues raised by the role of energy in social development fall into two broad categories: those that relate to the highly developed regions of the world and those that relate to the underdeveloped regions. Because industrialized nations now have the capability both for sustaining a modestly increasing population and for improving the average quality of life, it is likely that in the next century the per capita energy consumption in advanced countries will approach an equilibrium level.

For the underdeveloped nations, which include most of the world's population, the situation is quite different. The peoples of these nations are still primarily engaged in maintaining a minimum level of subsistence; they do not have available the power resources necessary for their transition to a literate, industrial, urban and advanced agricultural society. Such a transition will be significantly dependent on the availability of energy. It is sometimes suggested that because power production and energy consumption have harmful effects on the environment the use of energy must be arbitrarily limited. This implies the same type of social control as arbitrarily limiting the water supply, food production or population. Given the humane objective of providing the people of the world with a quality of life as high as man's ingenuity can develop, the essential role of energy must be accepted.

Within nature's limitations man has tremendous scope for planning energy utilization [*see illustration on these two pages*]. Some of the controlling factors that enter into energy policy depend on the voluntary decisions of the individual as well as on government actions that may restrict individual freedom. The questions of feasibility, both economic and technical, depend for their solution on the priority and magnitude of the effort applied. The time scale and costs for implementing decisions, or resolving issues, in all areas of energy management have both short-term and long-term consequences. There are so many variables that their arrangement into a "scenario" for the future becomes a matter of individual choice and a fascinating planning game. The intellectual complexity of the possible arrangements for the future can, however, be reduced to a limited number of basic policy questions that are more sociological than technical in nature.

The first set of questions has to do with the development of energy availability. These might be succinctly stated as follows. Whose resources should be utilized? Where should power be generated? Who shall receive the polluting effluents from such activities? These questions are particularly significant because fuel resources can be shipped all over the world by inexpensive ocean transport and electric power can be transmitted as needed over grids of continental size.

Our present approach to fuel sources has resulted in an international network for the tapping of the world's oil resources. Until World War II the U.S. was a net exporter of energy supplies. Today the Middle East and Africa are the major suppliers of oil, and they possess more than half of the world's fossil-fuel reserves. Thus the underdeveloped parts of the world are exporting their natural fuel resources to the developed parts of the world. Both western Europe and Japan are unique cases of highly industrialized areas almost completely dependent on the importation of oil from underdeveloped countries. The economics of this situation has been a prime factor in discouraging the U.S. from meeting its petroleum needs with oil shale and tar sands, which are available in very large amounts. Nevertheless, both the political and the sociological consequences of depending on foreign sources of supply make it likely that the oil shale and tar sands will be tapped even at high cost. Long-range planning to prepare for this technical development obviously has to be included in any national consideration of energy policies.

A current example of the knotty issues involved in separating the source of power from the user is found in the Four Corners power-region development in the U.S. The Four Corners embrace the region where Utah, New Mexico, Arizona and Colorado meet. It is a region with abundant reserves of low-sulfur coal, plentiful cooling water from the Colorado River and a low population density. The original plan was to build six coal-fired plants to provide electric power primarily for the large cities of Los Angeles, Las Vegas, Phoenix, Al-

	NATURAL LIMITATIONS
	100 – 1,000
RESOURCES	
FINITE FOSSIL-FUEL RESERVES	
URANIUM USAGE DEPENDS ON BREEDER	
CONTINUOUS SOURCES	
LIMITED EXCEPT FOR SOLAR	
ENVIRONMENTAL EFFECTS	
THERMODYNAMIC LIMIT ON CONVERSION EFFICIENCY	
REGIONAL CLIMATIC EFFECTS	
CO_2 PRODUCTION INEVITABLE FROM FOSSIL FUELS	

and technical feasibility and natural limitations, involve the fate of future generations.

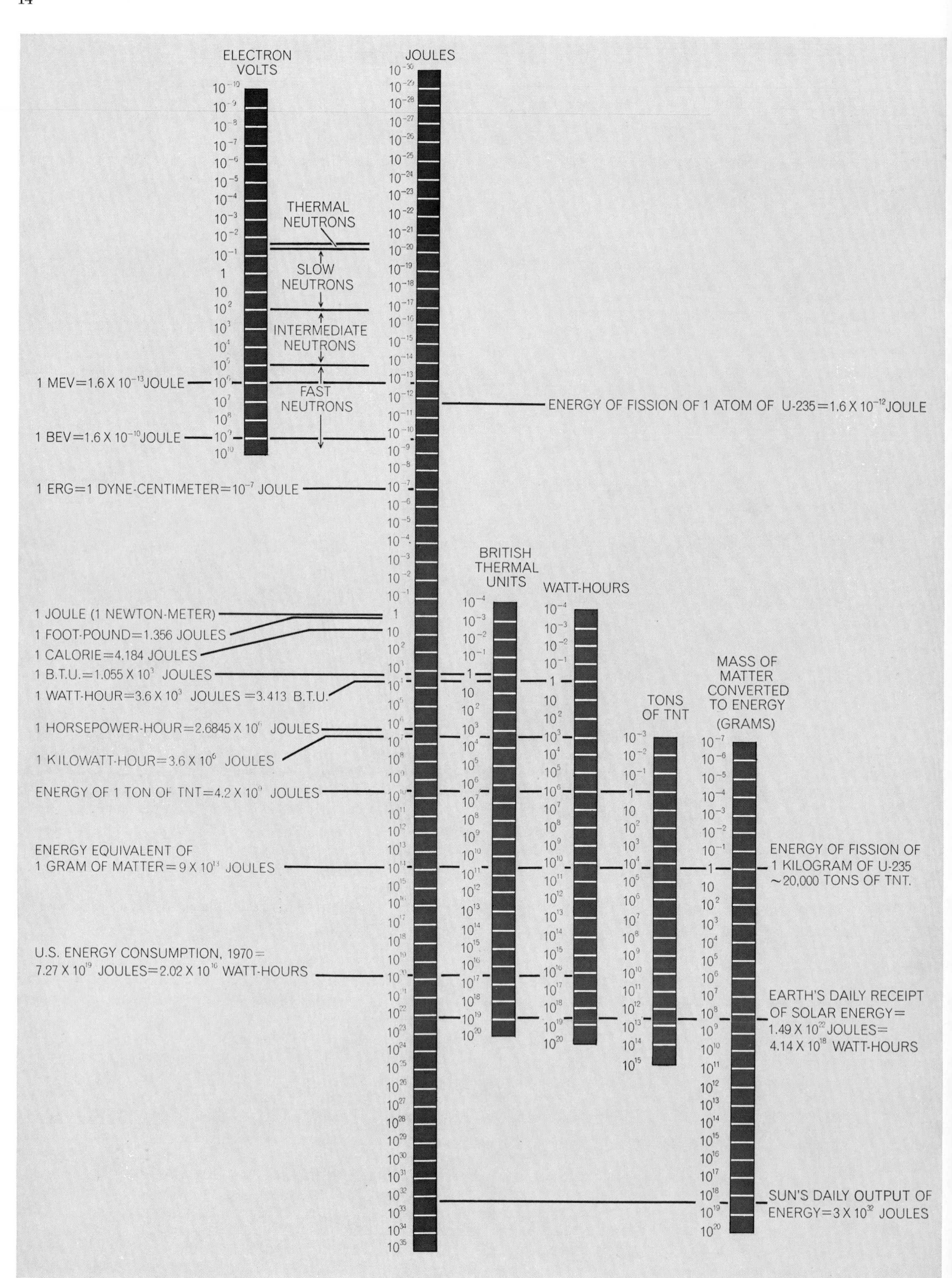
ELECTRON VOLTS
JOULES
THERMAL NEUTRONS
SLOW NEUTRONS
INTERMEDIATE NEUTRONS
FAST NEUTRONS
1 MEV=1.6 X 10^-13 JOULE
1 BEV=1.6 X 10^-10 JOULE
ENERGY OF FISSION OF 1 ATOM OF U-235=1.6 X 10^-12 JOULE
1 ERG=1 DYNE-CENTIMETER=10^-7 JOULE
BRITISH THERMAL UNITS
WATT-HOURS
1 JOULE (1 NEWTON-METER)
1 FOOT-POUND=1.356 JOULES
1 CALORIE=4.184 JOULES
1 B.T.U.=1.055 X 10^3 JOULES
1 WATT-HOUR=3.6 X 10^3 JOULES =3.413 B.T.U.
MASS OF MATTER CONVERTED TO ENERGY (GRAMS)
TONS OF TNT
1 HORSEPOWER-HOUR=2.6845 X 10^6 JOULES
1 KILOWATT-HOUR=3.6 X 10^6 JOULES
ENERGY OF 1 TON OF TNT=4.2 X 10^9 JOULES
ENERGY EQUIVALENT OF 1 GRAM OF MATTER=9 X 10^13 JOULES
ENERGY OF FISSION OF 1 KILOGRAM OF U-235 ~20,000 TONS OF TNT.
U.S. ENERGY CONSUMPTION, 1970= 7.27 X 10^19 JOULES=2.02 X 10^16 WATT-HOURS
EARTH'S DAILY RECEIPT OF SOLAR ENERGY= 1.49 X 10^22 JOULES= 4.14 X 10^18 WATT-HOURS
SUN'S DAILY OUTPUT OF ENERGY=3 X 10^32 JOULES

buquerque and other urban communities of the southwestern U.S.—all far distant from the Four Corners. In spite of the low population density of the Four Corners area, considerable protest has arisen over the environmental effects of intensive strip mining, the use of the Colorado for waste-heat disposal and a large-scale outpouring of stack effluents. Although some compromise of social benefits and penalties will presumably be reached to determine the acceptable levels of pollution from these plants, the Four Corners scheme epitomizes the kind of problem we can expect to encounter increasingly in planning large energy centers.

The issue of who gets the pollution, as contrasted with who gets the energy, is not only one of geographic distribution but also one of time. For example, if as a result of the rapid increase in strip-mining for coal, the acid drainage and soil erosion disrupt ecosystems over a large region, it may take decades to repair the damage in spite of the coal mining company's genuine effort to restore the local area to a semblance of its original condition. This generation of energy users will have been long gone when succeeding generations face the problem and the cost of repairing the damage of such ecological degradation. Other long-term and long-delayed problems may be associated with the effluents released by using power to do useful work. The penalties imposed on future generations are the result of social choices made today.

Another category of choices relate to the way we use energy after allocating the fraction necessary for doing useful work. Our society provides many options in which energy is used for recreation, environmental conditioning, communication and entertainment. An automobile tour of a country, a powerboat cruise, an airplane vacation trip all represent energy consumption subject to individual choice and taste. They also represent choices that produce effluents with some effect on the environment. Are we prepared to limit this freedom of choice, which implies the freedom to pollute? The answer will require a careful balancing of values.

Most significant is the allocation of our national resources, manpower and technology to the improvement of the physical environment as compared with other needs. If we did not count the cost, there is little doubt we could so reduce the effluents from the utilization of energy that their effects on public health would be truly negligible. The cost, however, might be excessively large. Thus one must ask if there is an intermediate level of control that is acceptable for comfort, health and aesthetics. The continuous exposure of man to many natural pollutants (uninfluenced by man) is great enough so that there may not be much justification for reducing the pollutants of energy conversion much below the natural background levels. Since even the wealthy U.S. cannot satisfy all the demands on its resources, the level of pollution control seems bound to emerge as a major factor in the debate over national priorities. For example, it may be much more important to allocate resources to improving public health services rather than to use that same sum to marginally reduce environmental pollution. It is unfortunately true that in a pluralistic society the value systems and priorities differ among the society's sectors. The groups seeking aesthetic and environmental improvement may be a minority compared with the much larger number seeking basic material improvements. In the energy field a decision on a national level concerning the energy system may be a determining factor in shaping the framework of our society for some generations to come.

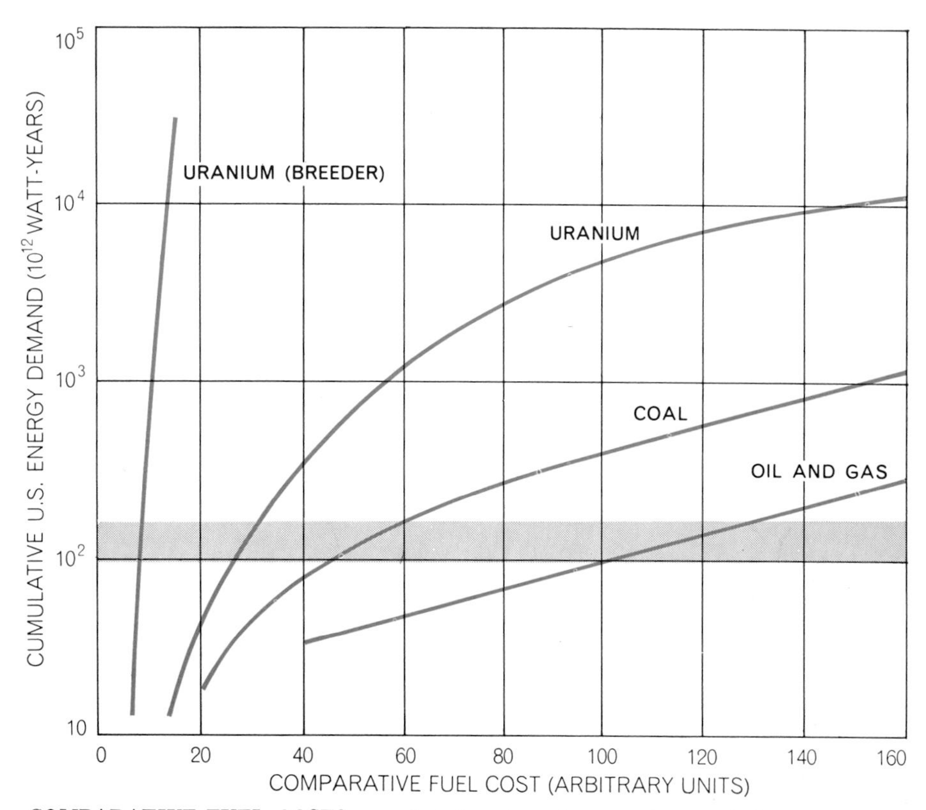

COMPARATIVE FUEL COSTS are plotted against the nation's cumulative demand for energy. The horizontal band covers the range in probable demand from 1960 to 2000. Uranium used in present-day nonbreeder reactors is already cheaper than fossil fuels. The fast-breeder reactor should hold fuel costs essentially constant far beyond the year 2000.

Perhaps the most fundamental question of national policy is how we should allocate our present resources for the benefit of future generations. The development of new speculative energy resources is an investment for the future, not a means of remedying the problems of today. It is equally clear that the quality of life of the peoples of the world depends on the availability *now* of large amounts of low-cost energy in useful form. This being so, we must emphasize an orderly development of the resources available to us with present technology, and these are primarily power plants based on fossil fuels and nuclear fission.

ENERGY UNITS and conversion factors are presented on the opposite page. Physicists find it convenient to use electron volts, ergs and joules. Biologists and nutritionists think in calories. Engineers deal in British thermal units and watt-hours. Since Hiroshima energy release is commonly expressed in tons of TNT. It is less often observed that a ton of ordinary coal contains three times as much energy as a ton of TNT. The illustration is based on one that appears in *The New College Physics: A Spiral Approach*, by Albert V. Baez.

2

Energy in the Universe

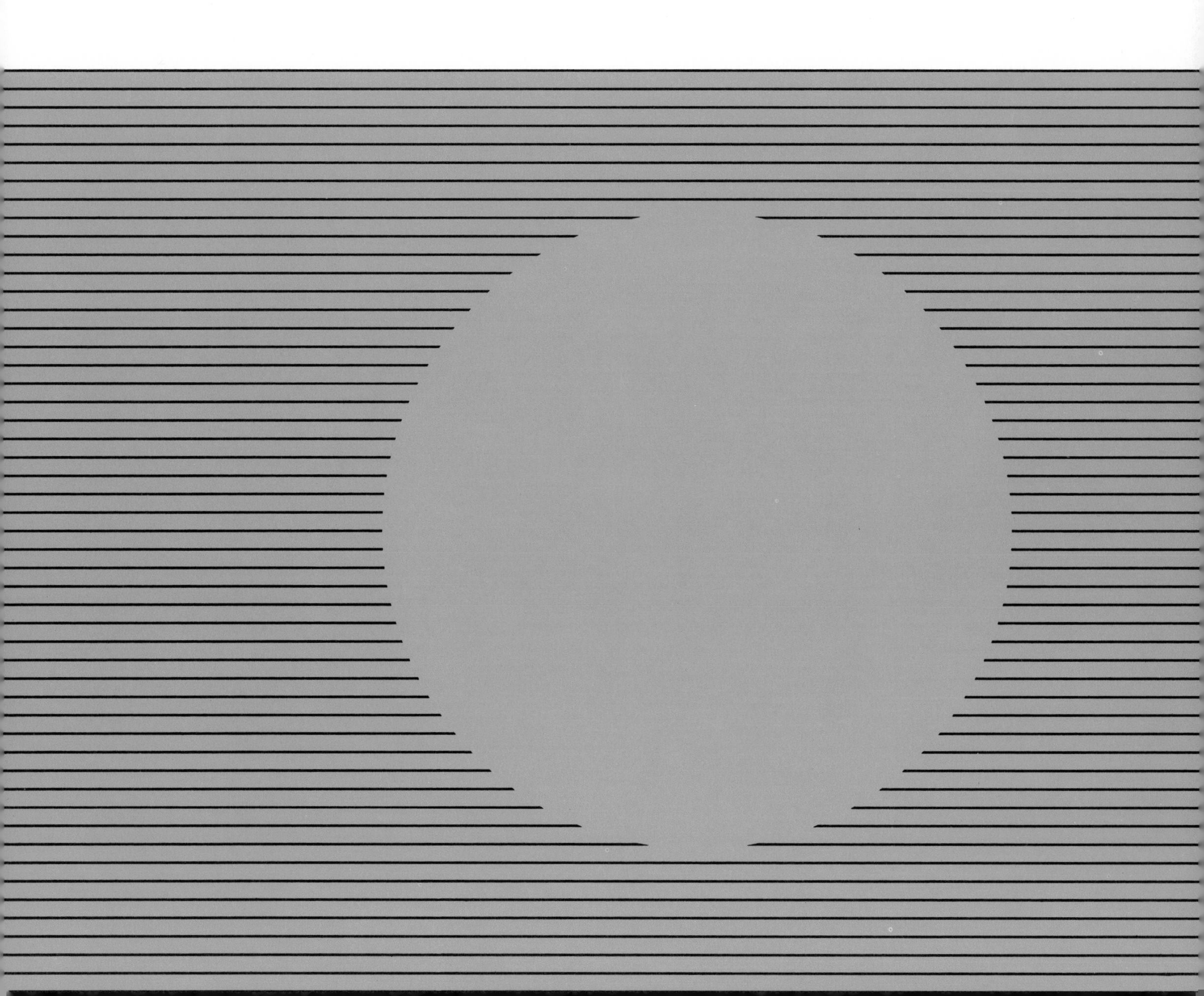

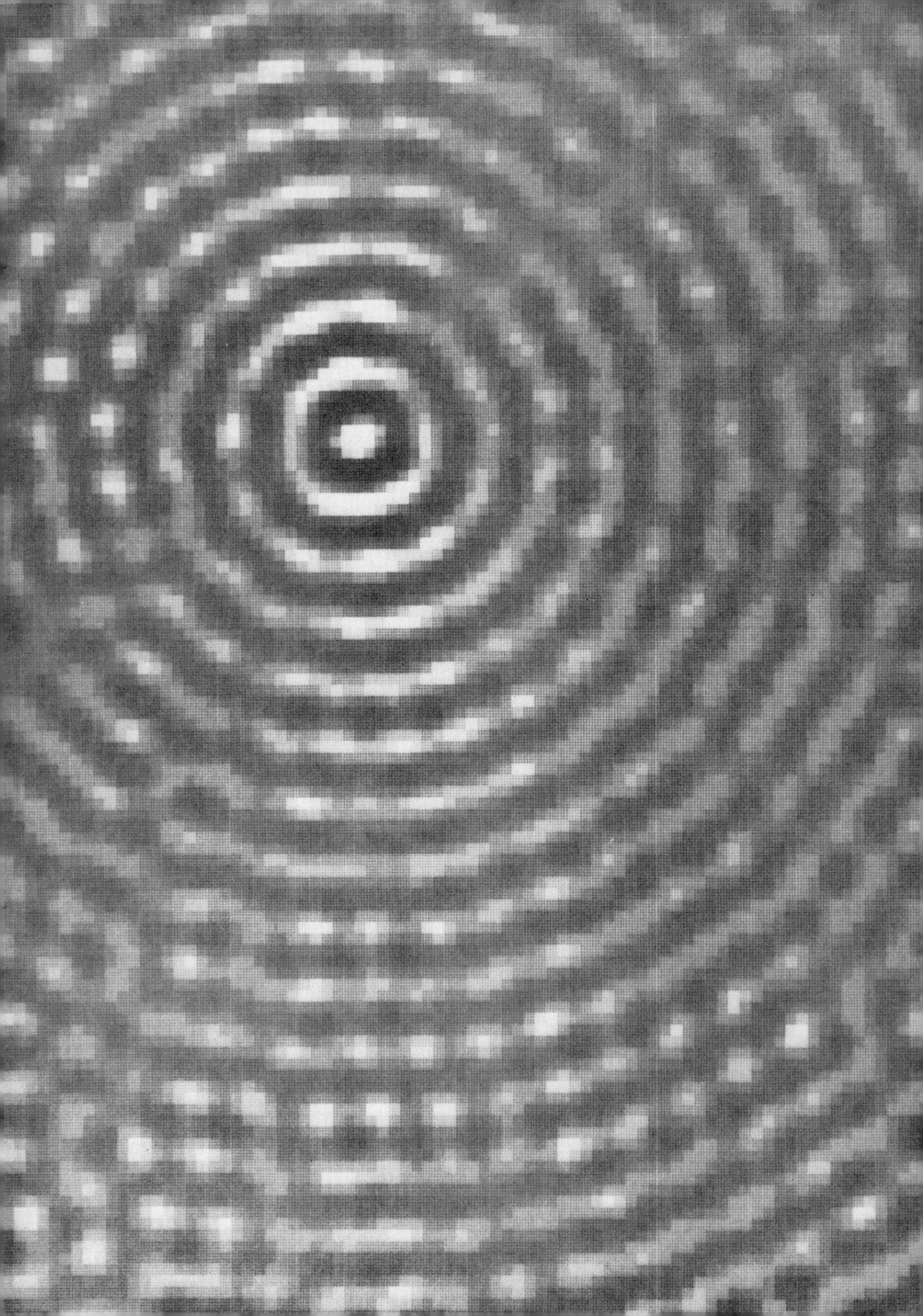

Energy in the Universe

FREEMAN J. DYSON

The energy flows on the earth are embedded in the energy flows in the universe. A delicate balance among gravitation, nuclear reactions and radiation keeps the energy from flowing too fast

Man has no Body distinct from his Soul;
for that called Body is a portion of Soul
discern'd by the five Senses, the chief
inlets of Soul in this age.

Energy is the only life and is from the Body;
and Reason is the bound or outward circumference of Energy.

Energy is Eternal Delight.

—William Blake,
The Marriage of Heaven and Hell, 1793

One need not be a poet or a mystic to find Blake's definition of energy more satisfying than the definitions given in textbooks on physics. Even within the framework of physical science energy has a transcendent quality. On many occasions when revolutions in thought have demolished old sciences and created new ones, the concept of energy has proved to be more valid and durable than the definitions in which it was embodied. In Newtonian mechanics energy was defined as a property of moving masses. In the 19th century energy became a unifying principle in the construction of three new sciences: thermodynamics, quantitative chemistry and electromagnetism. In the 20th century energy again appeared in fresh disguise, playing basic and unexpected roles in the twin intellectual revolutions that led to relativity theory and quantum theory. In the special theory of relativity Einstein's equation $E = mc^2$, identifying energy with mass, threw a new light on our view of the astronomical universe, a light whose brilliance no amount of journalistic exaggeration has been able to obscure. And in quantum mechanics Planck's equation $E = h\nu$, restricting the energy carried by any oscillation to a constant multiple of its frequencies, transformed in an even more fundamental way our view of the subatomic universe. It is unlikely that the metamorphoses of the concept of energy, and its fertility in giving birth to new sciences, are yet at an end. We do not know how the scientists of the next century will define energy or in what strange jargon they will discuss it. But no matter what language the physicists use they will not come into contradiction with Blake. Energy will remain in some sense the lord and giver of life, a reality transcending our mathematical descriptions. Its nature lies at the heart of the mystery of our existence as animate beings in an inanimate universe.

The purpose of this article is to give an account of the movement of energy in the astronomical world, insofar as we understand it. I shall discuss the genesis of the various kinds of energy that are observed on the earth and in the sky, and the processes by which energy is channeled in the evolution of stars and galaxies. This overall view of the sources and flow of energy in the cosmos is intended to put in perspective the articles that follow, which deal with the problems of the use of energy by mankind on the earth. In looking to our local energy resources it is well to consider how we fit into the larger scheme of things. Ultimately what we can do here on the earth will be limited by the same laws that govern the economy of astronomical energy sources. The converse of this statement may also be true. It would not be surprising if it should turn out that the origin and destiny of the energy in the universe cannot be completely understood in isolation from the phenomena of life and consciousness. As we look out into space we see no sign that life has intervened to control events anywhere except precariously on our own planet. Everywhere else the universe appears to be mindlessly burning up its reserves of energy, inexorably drifting toward the state of final quiescence described imaginatively by Olaf Stapledon: "Presently nothing was left in the whole cosmos but darkness and the dark whiffs of dust that once were galaxies." It is conceivable, however, that life may have a larger role to play than we have yet imagined. Life may succeed against all the odds in molding the universe to its own purposes. And the design of the inanimate universe may not be as detached from the potentialities of life and intelligence as scientists of the 20th century have tended to suppose.

The cosmos contains energy in various forms, for example gravitation, heat, light and nuclear energy. Chemical energy, the form that plays the major role in present-day human activities, counts for very little in the universe as a whole.

CELESTIAL ENERGY SOURCE is represented by the computer-generated display on the opposite page, which is based on data gathered by means of a rocket-borne X-ray detection device. The display, known as a correlation map, was used to locate with high precision a strong X-ray source (designated GX 5–1) in the constellation Sagittarius near the direction of the galactic center. The experiment was carried out by a team of investigators from the Massachusetts Institute of Technology; the details of the experimental procedure are described in the September 1970 issue of *The Astrophysical Journal*. The mechanism responsible for the large energy fluxes emanating from such sources is unknown, but it is believed to play an important role in the overall energy flow of the universe.

In the universe the predominant form of energy is gravitational. Every mass spread out in space possesses gravitational energy, which can be released or converted into light and heat by letting the mass fall together. For any sufficiently large mass this form of energy outweighs all others.

The laws of thermodynamics decree that each quantity of energy has a characteristic quality called entropy associated with it. The entropy measures the degree of disorder associated with the energy. Energy must always flow in such a direction that the entropy increases. Thus we can arrange the different forms of energy in an "order of merit," the highest form being the one with the least disorder or entropy [*see illustration below*]. Energy of a higher form can be degraded into a lower form, but a lower form can never be wholly converted back into a higher form. The basic fact determining the direction of energy flow in the universe is that gravitational energy is not only predominant in quantity but also highest in quality. Gravitation carries no entropy and stands first in the order of merit. It is for this reason that a hydroelectric power station converting the gravitational energy of water to electricity can have an efficiency close to 100 percent, which no chemical or nuclear power station can approach. In the universe as a whole the main theme of energy flow is the gravitational contraction of massive objects, the gravitational energy released in contraction being converted into energy of motion, light and heat. The flow of water from a reservoir to a turbine situated a little closer to the center of the earth is in essence a controlled gravitational contraction of the earth, only on a more modest scale than astronomers are accustomed to consider. The universe evolves by the gravitational contraction of objects of all sizes, from clusters of galaxies to planets.

When one views the universe in broad outline in this way, a set of paradoxical questions at once arises. Since thermodynamics favors the degradation of gravitational energy to other forms, how does it happen that the gravitational energy of the universe is still predominant after 10 billion years of cosmic evolution? Since large masses are unstable against gravitational collapse, why did they not all collapse long ago and convert their gravitational energy into heat and light in a quick display of cosmic fireworks? Since the universe is on a one-way slide toward a state of final death in which energy is maximally degraded, how does it manage, like King Charles, to take such an unconscionably long time a-dying? These questions are not easy to answer. The further one goes in answering them, the more remarkable and paradoxical becomes the apparent stability of the cosmos. It turns out that the universe as we know it survives not by any inherent stability but by a succession of seemingly accidental "hangups." By a hangup I mean an obstacle, usually arising from some quantitative feature of the design of the universe, that arrests the normal processes of degradation of energy. Psychological hangups are generally supposed to be bad for us, but cosmological hangups are absolutely necessary for our existence.

FORM OF ENERGY	ENTROPY PER UNIT ENERGY
GRAVITATION	0
ENERGY OF ROTATION	0
ENERGY OF ORBITAL MOTION	0
NUCLEAR REACTIONS	10^{-6}
INTERNAL HEAT OF STARS	10^{-3}
SUNLIGHT	1
CHEMICAL REACTIONS	1–10
TERRESTRIAL WASTE HEAT	10–100
COSMIC MICROWAVE RADIATION	10^4

"ORDER OF MERIT" of the major forms of energy in the universe ranks the various energy forms roughly according to their associated entropy per unit energy, expressed in units of inverse electron volts. The entropy, which measures the degree of disorder associated with a particular form of energy, varies approximately inversely with the temperature associated with that energy form. In the cases of gravitation, rotation and orbital motion there is no associated temperature and hence the entropy is zero. Energy generally flows from higher levels to lower levels in the table, that is, in such a direction that the entropy increases. The cosmic microwave background radiation appears to be an ultimate heat sink; no way is known in which this energy could be further degraded or converted into any other form. The universe survives not by any inherent stability but by a succession of seemingly accidental "hangups," or obstacles, usually arising from some quantitative feature of the design of the universe, that act to arrest the normal processes of the degradation of energy.

The first and most basic hangup built into the architecture of the universe is the size hangup. A naïve person looking at the cosmos has the impression that the whole thing is extravagantly, even irrelevantly, large. This extravagant size is our primary protection against a variety of catastrophes. If a volume of space is filled with matter with an average density d, the matter cannot collapse gravitationally in a time shorter than the "free-fall time" t, which is the time it would take to fall together in the absence of any other hangups. The formula relating d with t is $Gdt^2 = 1$, where G is the constant in Newton's law of gravitation. The effect of this formula is that when we have an extravagantly small density d, and therefore an extravagantly big volume of space, the free-fall time t can become so long that gravitational collapse is postponed to a remote future.

If we take for d the average density of mass in the visible universe, which works out to about one atom per cubic meter, the free-fall time is about 100 billion years. This is longer than the probable age of the universe (10 billion years), but only by a factor of 10. If the matter in the universe were not spread out with such an exceedingly low density, the free-fall time would already have ended and our remote ancestors would long ago have been engulfed and incinerated in a universal cosmic collapse.

The matter inside our own galaxy has an average density about a million times higher than that of the universe as a whole. The free-fall time for the galaxy is therefore about 100 million years. Within the time span of life on the earth the galaxy is not preserved from gravitational collapse by size alone. Our survival requires other hangups besides the hangup of size.

Another form of degradation of gravitational energy, one less drastic than gravitational collapse, would be the disruption of the solar system by close encounters or collisions with other stars. Such a degradation of the orbital motions of the earth and planets would be just as fatal to our existence as a complete collapse. We have escaped this catastrophe only because the distances between stars in our galaxy are also extravagantly large. Again a calculation shows that our galaxy is barely large enough to make the damaging encounters unlikely. So even within our galaxy

SIZE HANGUP, the first and most basic hangup that is built into the architecture of the universe, is symbolized by this photograph of a large cluster of galaxies in the constellation Hercules. A cluster may contain anywhere from two galaxies to several thousand. It typically occupies approximately 10^{20} cubic light-years of space and maintains an average distance between galaxies of about a million light-years. It is the extravagantly large volume of space that is the primary protection against a variety of cosmic catastrophes. By making the "free-fall time" of the universe so long, for example, the size hangup postpones the ultimate gravitational collapse of the universe to a remote future. The photograph was made with the 200-inch Hale reflecting telescope on Palomar Mountain.

the size hangup is necessary to our preservation, although it is not by itself sufficient.

The second on the list of hangups is the spin hangup. An extended object cannot collapse gravitationally if it is spinning rapidly. Instead of collapsing, the outer parts of the object settle into stationary orbits revolving around the inner parts. Our galaxy as a whole is preserved by this hangup, and the earth is preserved by it from collapsing into the sun. Without the spin hangup no planetary system could have been formed at the time the sun condensed out of the interstellar gas.

The spin hangup has produced ordered structures with an impressive appearance of permanence, not only galaxies and planetary systems but also double stars and the rings of Saturn. None of these structures is truly permanent. Given sufficient time all will be degraded by slow processes of internal energy dissipation or by random encounters with other objects in the universe. The solar system seems at first to be a perfect perpetual motion machine, but in reality its longevity is dependent on the combined action of the spin hangup and the size hangup.

The third hangup is the thermonuclear hangup. This hangup arises from the fact that hydrogen "burns" to form helium when it is heated and compressed. The thermonuclear burning (actually fusion reactions between hydrogen nuclei) releases energy, which opposes any further compression. As a result any object such as a star that contains a large proportion of hydrogen is unable to collapse gravitationally beyond a certain point until the hydrogen is all burned up. For example, the sun has been stuck on the thermonuclear hangup for 4.5 billion years and will take about another five billion years to burn up its hydrogen before its gravitational contraction can be resumed [*see top illustration on page 24*]. Ultimately the supply of nuclear energy in the universe is only a small fraction of the supply of gravitational energy. But the nuclear energy acts as a delicately adjusted regulator, postponing the violent phases of gravitational collapse and allowing stars to shine peacefully for billions of years.

There is good evidence that the universe began its existence with all the matter in the form of hydrogen, with perhaps some admixture of helium but few traces of heavier elements. The evidence comes from the spectra of stars moving in our galaxy with very high velocities with respect to the sun. The high velocities mean that these stars do not take part in the general rotation of the galaxy. They are moving in orbits that are oblique to the plane of the galaxy, and therefore their velocity and the sun's combine to give a relative velocity of the order of hundreds of kilometers per second. Such a velocity is in contrast to that of common stars, which orbit with the sun in the central plane of the galaxy and show relative velocities of the order of

SPIN HANGUP is exemplified by this photograph of a typical galaxy of the "open spiral" type. Galaxies, planetary systems, double stars and the rings of Saturn are among the celestial objects that are spared temporarily from the inevitable gravitational collapse by the spin hangup. This particular galaxy, designated M 101, illustrates the mechanism of star formation that is probably still at work in our own galaxy. Each spiral arm consists of clumps of bright, newly formed stars left behind by the passage of a rotating hydrodynamic-gravitational wave in the galactic disk. Photograph was also made by the 200-inch telescope on Palomar Mountain.

tens of kilometers per second. The high-velocity stars form a "halo," or spherical cloud, that is bisected by the rotating galactic disk, which contains the bulk of the ordinary stars [*see top illustration at right*].

The obvious explanation of this state of affairs is that the high-velocity stars are the oldest. They condensed out of the primeval galaxy while it was still in a state of free fall, before it encountered the spin hangup. After the spin hangup the galaxy settled down into a disk, and the ordinary stars were formed in orbits within the disk, where they have remained ever since. This picture of the history of the galaxy is dramatically confirmed by the spectroscopic evidence [*see bottom illustration at right*]. The spectra of the extreme high-velocity stars show extremely weak absorption lines for all the elements except hydrogen. These stars are evidently composed of less than a tenth—sometimes less than a hundredth—as much of the common elements carbon, oxygen and iron as we find in the sun. Such major deficiencies of the common elements are almost never found in low-velocity stars. Since hydrogen burns to make carbon and iron, but carbon and iron cannot burn to make hydrogen, the objects with the least contamination of hydrogen by heavier elements must be the oldest. We can still see a few stars in our neighborhood dating back to a time so early that the contamination by heavier elements was close to zero.

The discovery that the universe was originally composed of rather pure hydrogen implies that the thermonuclear hangup is a universal phenomenon. Every mass large enough to be capable of gravitational collapse must inescapably pass through a prolonged hydrogen-burning phase. The only objects exempt from this rule are masses of planetary size or smaller, in which gravitational contraction is halted by the mechanical incompressibility of the material before the ignition point of thermonuclear reactions is reached. The preponderance of hydrogen in the universe ensures that our night sky is filled with well-behaved stars like our own sun, placidly pouring out their energy for the benefit of any attendant life-forms and giving to the celestial sphere its historic attribute of serene immobility. It is only by virtue of the thermonuclear hangup that the heavens have appeared to be immobile. We now know that in corners of the universe other than our own violent events are the rule rather than the exception. The prevalence of catastrophic outbursts of energy was revealed to us through the rapid

TWO STELLAR POPULATIONS are characteristically present in spiral galaxies, the older stars forming a roughly spherical halo cloud and the younger stars forming a comparatively thin central disk. This photograph of the spiral galaxy M 104, seen almost edge on, gives a particularly clear view of the two types of stellar population. In our own galaxy the arrangement is the same but the proportions are different. If our galaxy were seen this way, the disk would look much brighter than that of M 104, the halo much fainter.

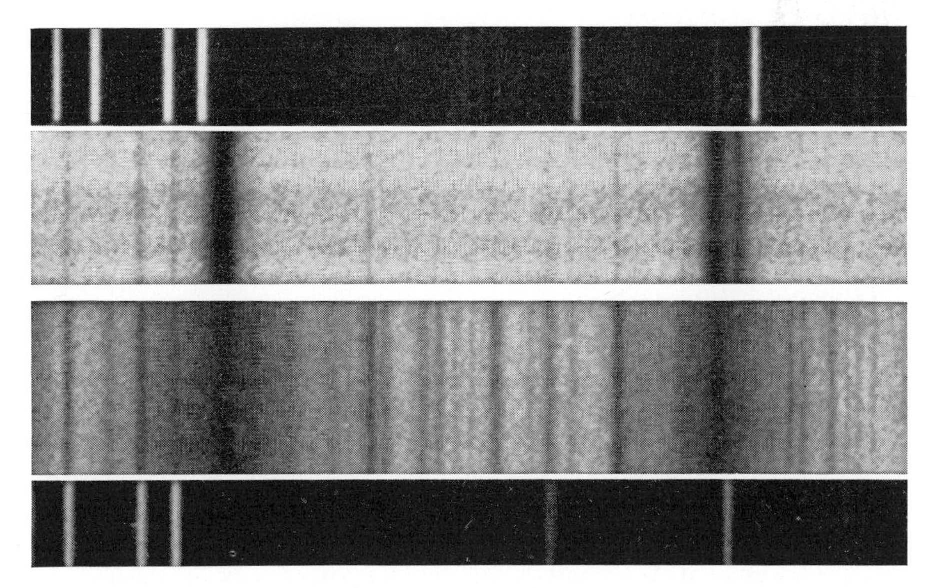

SPECTROGRAMS of two stars in our galaxy, a high-velocity halo star (*second from top*) and a normal disk star (*third from top*) of approximately the same spectral type, were made with the 120-inch telescope at the Lick Observatory. The spectrum of the high-velocity star, designated HD 140283, shows comparatively weak absorption lines for all the elements except hydrogen. The spectrum of the normal star, our own sun, contains numerous lines associated with the heavier elements, particularly carbon and iron. Since hydrogen burns to make carbon and iron but carbon and iron cannot burn to make hydrogen, the objects in our galaxy with the least contamination of hydrogen by heavier elements must be the oldest. Spectra at top and bottom provide bright lines for reference purposes.

STAGE	DURATION	RANGE OF OUTPUT OF ENERGY
1. GRAVITATIONAL CONTRACTION	143 MILLION YEARS	INITIALLY 600, DECREASING RAPIDLY TO .7
2. HYDROGEN BURNING	10.3 BILLION YEARS	INITIALLY .7, INCREASING SLOWLY TO 3
3. RESUMED GRAVITATIONAL CONTRACTION OF CORE	500 MILLION YEARS	3–10
4. HELIUM AND CARBON BURNING	500 MILLION YEARS	10–1,000, FLUCTUATING IN COMPLICATED FASHION
5. FINAL GRAVITATIONAL CONTRACTION	13 MILLION YEARS	1,000–.01
6. WHITE-DWARF PHASE	INFINITE	.01, COOLING SLOWLY TO ZERO

EFFECT OF THERMONUCLEAR HANGUP on the life of the sun is evident in this table, which gives the sun's energy output at various stages in its evolution in units of the present solar luminosity (2.10^{33} ergs per second). Only the first three stages are known well enough to be accurately computed. The details of stages *4* and *5* are uncertain because the mechanisms of convective instability in the sun's interior and of mass loss at the surface are not completely understood. During stage *4* the sun will probably pass through a "red giant" phase, and during stage *5* through a "planetary nebula" phase. What is certain is that in stages *4* and *5* the energy output will be high and the duration short compared with the energy output and duration of stage *2*. If it were not for the thermonuclear hangup of stage *2*, the sun would have squandered all its energy and reached stage *6* long ago, probably in less than a billion years. As matters stand, the sun is only halfway through stage *2*.

progress of radio astronomy over the past 30 years. These outbursts are still poorly understood, but it seems likely that they occur in regions of the universe where the thermonuclear hangup has been brought to an end by the exhaustion of hydrogen.

It may seem paradoxical that the thermonuclear hangup has such benign and pacifying effects on extraterrestrial affairs in view of the fact that, so far at least, our terrestrial thermonuclear devices are neither peaceful nor particularly benign. Why does the sun burn its hydrogen gently for billions of years instead of blowing up like a bomb? To answer this question it is necessary to invoke yet another hangup.

The crucial difference between the sun and a bomb is that the sun contains ordinary hydrogen with only a trace of the heavy hydrogen isotopes deuterium and tritium, whereas the bomb is made mainly of heavy hydrogen. Heavy hydrogen can burn explosively by strong nuclear interactions, but ordinary hydrogen can react with itself only by the weak-interaction process. In this process two hydrogen nuclei (protons) fuse to form a deuteron (a proton and a neutron) plus a positron and a neutrino. The proton-proton reaction proceeds about 10^{18} times more slowly than a strong nuclear reaction at the same density and temperature. It is this weak-interaction hangup that makes ordinary hydrogen useless to us as a terrestrial source of energy. The hangup is essential to our existence, however, in at least three ways. First, without this hangup we would not have a sufficiently long-lived and stable sun. Second, without it the ocean would be an excellent thermonuclear high explosive and would constitute a perennial temptation to builders of "doomsday machines." Third and most important, without the weak-interaction hangup it is unlikely that any appreciable quantity of hydrogen would have survived the initial hot, dense phase of the evolution of the universe. Essentially all the matter in the universe would have been burned to helium before the first galaxies started to condense, and no normally long-lived stars would have had a chance to be born.

If one looks in greater detail at the theoretical reasons for the existence of the weak-interaction hangup, our salvation seems even more providential. The hangup depends decisively on the nonexistence of an isotope of helium with a mass number of 2, the nucleus of which would consist of two protons and no neutrons. If helium 2 existed, the proton-proton reaction would yield a helium-2 nucleus plus a photon, and the helium-2 nucleus would in turn spontaneously decay into a deuteron, a positron and a neutrino. The first reaction being strong, the hydrogen would burn fast to produce helium 2. The subsequent weak decay of the helium 2 would not limit the rate of burning. It happens that there does exist a well-observed state of the helium-2 nucleus, but the state is unbound by about half a million volts. The nuclear force between two protons is attractive and of the order of 20 million volts, but it just barely fails to produce a bound state. If

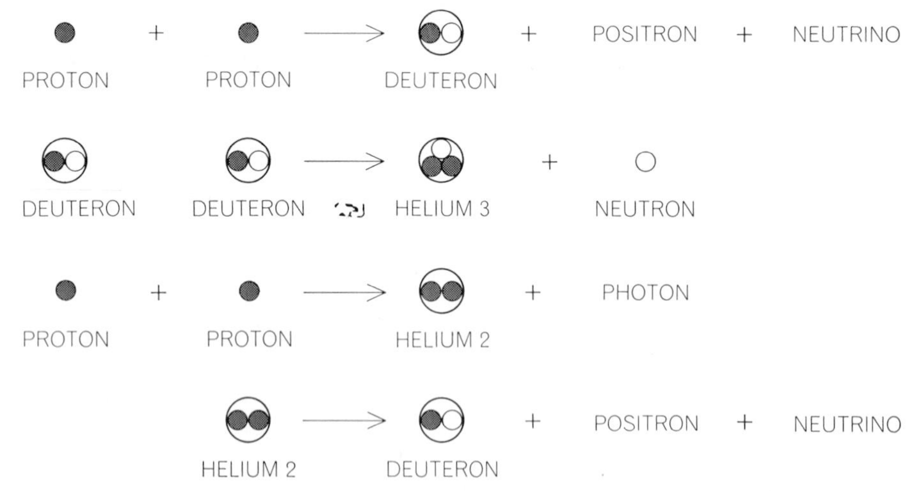

SOME FUSION REACTIONS discussed in this article are depicted schematically here. In the sun (*top*) ordinary hydrogen nuclei (protons) fuse to form a deuteron (a proton and a neutron) plus a positron and a neutrino. In a thermonuclear bomb (*second from top*) two heavy hydrogen isotopes, in this case both deuterons, fuse by the strong interaction process to form a helium-3 nucleus plus a neutron. The proton-proton reaction proceeds about 10^{18} times more slowly than the corresponding deuteron-deuteron reaction. If a helium-2 nucleus could exist, the proton-proton reaction would yield a helium-2 nucleus plus a photon (*third from top*), and the helium-2 nucleus would in turn spontaneously decay into a deuteron, a positron and a neutrino (*fourth from top*). As a consequence there would be no weak-interaction hangup, and essentially all of the hydrogen existing in the universe would have been burned to helium even before the first galaxies had started to condense.

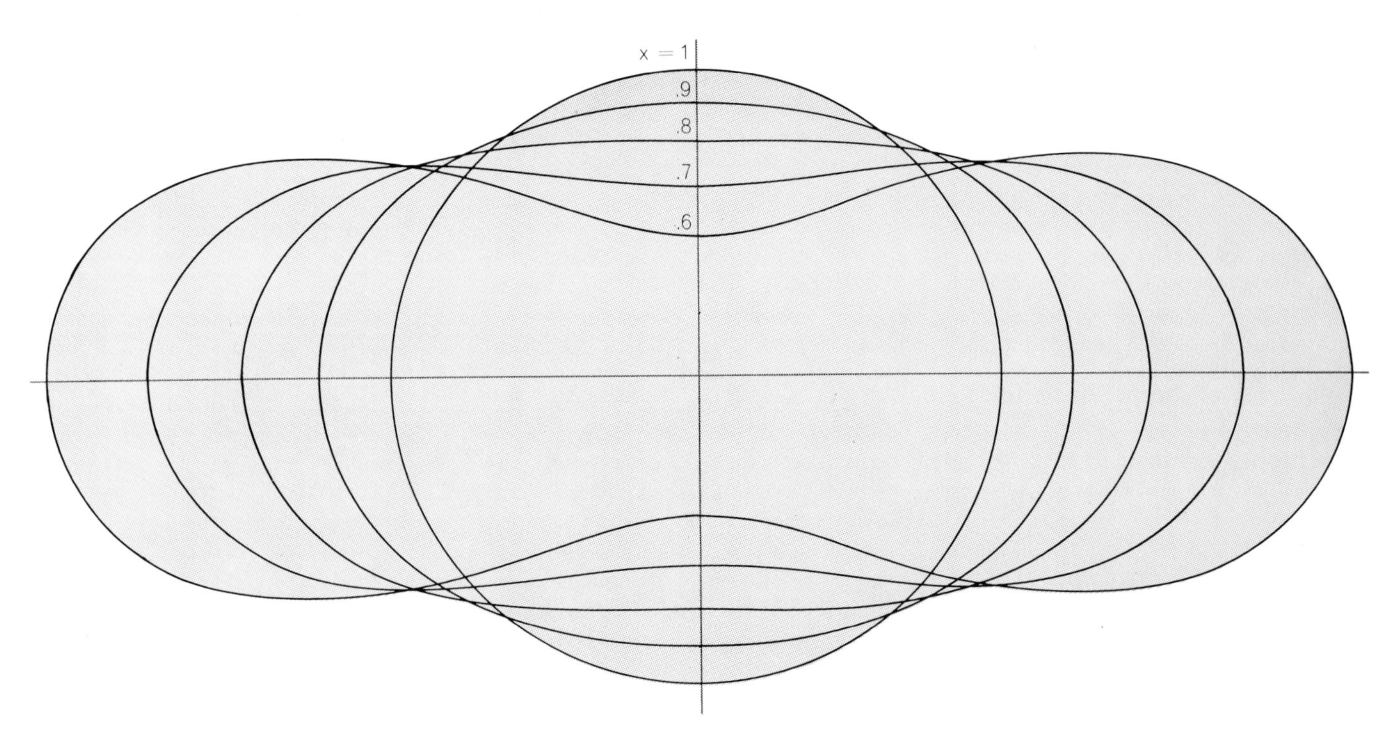

SURFACE-TENSION HANGUP has enabled fissionable nuclei such as uranium to survive in the earth's crust for aeons. Before these nuclei can split spontaneously their surface must be stretched into a nonspherical shape, and this stretching is opposed by an extremely powerful force of surface tension. This diagram shows the shapes of various nuclei when they go "over the hump" during fission; the shapes were computed according to the liquid-drop model of the nucleus. The nuclei are labeled by a parameter *x*, which is the ratio of electrostatic energy to surface tension. Nuclei from thorium to plutonium all have values of *x* between .7 and .8. The larger *x* is, the more unstable the nucleus is and the smaller the deformation required before fission occurs.

the force were a few percent stronger, there would be no weak-interaction hangup.

I have discussed four hangups: size, spin, thermonuclear and weak-interaction. The catalogue is by no means complete. There is an important class of transport or opacity hangups, which arise because the transport of energy by conduction or radiation from the hot interior of the earth or the sun to the cooler surface takes billions of years to complete. It is the transport hangup that keeps the earth fluid and geologically active, giving us such phenomena as continental drift, earthquakes, volcanoes and mountain uplift. All these processes derive their energy from the original gravitational condensation of the earth four billion years ago, supplemented by a modest energy input from subsequent radioactivity.

Last on my list is a special surface-tension hangup that has enabled the fissionable nuclei of uranium and thorium to survive in the earth's crust until we are ready to use them. These nuclei are unstable against spontaneous fission. They contain so much positive charge and so much electrostatic energy that they are ready to fly apart at the slightest provocation. Before they can fly apart, however, their surface must be stretched into a nonspherical shape, and this stretching is opposed by an extremely powerful force of surface tension. A nucleus is kept spherical in exactly the same way a droplet of rain is kept spherical by the surface tension of water, except that the nucleus has a tension about 10^{18} times as strong as that of the raindrop. In spite of this surface tension a nucleus of uranium 238 does occasionally fission spontaneously, and the rate of the fissioning can be measured. Nonetheless, the hangup is so effective that less than one in a million of the earth's uranium nuclei has disappeared in this way during the whole of geological history.

No hangup can last forever. There are times and places in the universe at which the flow of energy breaks through all hangups. Then rapid and violent transformations occur, of whose nature we are still ignorant. Historically it was physicists and not astronomers who recorded the first evidence that the universe is not everywhere as quiescent as traditional astronomy had pictured it. The physicist Victor Hess discovered 60 years ago that even our quiet corner of the galaxy is filled with a uniform cloud of the extremely energetic particles now called cosmic rays. We still do not know in detail where these particles come from, but we do know that they represent an important channel in the overall energy flow of the universe. They carry on the average about as much energy as starlight.

The cosmic rays must certainly originate in catastrophic processes. Various attempts to explain them as by-products of familiar astronomical objects have proved quantitatively inadequate. In the past 30 years half a dozen strange new types of object have been discovered, each of which is violent and enigmatic enough to be a plausible parent of cosmic rays. These include the supernovas (exploding stars), the radio galaxies (giant clouds of enormously energetic electrons emerging from galaxies), the Seyfert galaxies (galaxies with intensely bright and turbulent nuclei), the X-ray sources, the quasars and the pulsars. All these objects are inconspicuous only because they are extremely distant from us. And once again only the size hangup —the vastness of the interstellar spaces—has diluted the cosmic rays enough to save us from being fried or at least sterilized by them. If sheer distance had not effectively isolated the quiet regions of the universe from the noisy ones, no type of biological evolution would have been possible.

The longest-observed and least mysterious of the violent objects are the supernovas. These appear to be ordinary stars, rather more massive than the sun, that have burned up their hydrogen and passed into a phase of gravitational collapse. In various ways the rapid release of gravitational energy can cause the star to explode. There may in some cases be a true thermonuclear detonation, with the core of the star, composed mainly of carbon and oxygen, burning instantaneously to iron. In other cases the collapse may cause the star to spin so rapidly that hydrodynamic instability disrupts it. A third possibility is that a spinning magnetic field becomes so intensified by gravitational collapse that it can drive off the surface of the star at high velocity. Probably several different kinds of supernova exist, each with a different mechanism of energy transfer. In all cases the basic process must be a gravitational collapse of the core of the star. By one means or another some fraction of the gravitational energy released by the collapse is transferred outward and causes the outer layers of the star to explode. The outward-moving energy appears partly as visible light, partly as the energy of motion of the debris and partly as the energy of cosmic rays. In addition a small fraction of the energy may be converted into the nuclear energy of the unstable nuclear species thorium and uranium, and small amounts of these elements may be injected by the explosion into the interstellar gas. As far as we know no other mechanism can create the special conditions required for the production of fissionable nuclei.

We have firm evidence that a locally violent environment existed in our galaxy immediately before the birth of the solar system. It is likely that the violence and the origin of the sun and the earth were part of the same sequence of events. The evidence for violence is the existence in certain very ancient meteorites of xenon gas with an isotopic composition characteristic of the products of spontaneous fission of the nucleus plutonium 244. Supporting evidence is provided by radiation damage in the form of fission-fragment tracks that can be made visible by etching in pieces of other meteorites [*see illustration below*]. The meteorites do not contain enough uranium or thorium to account for either the xenon or the fission tracks. Plutonium 244, although it is the longest-lived isotope of plutonium, has a half-life of only 80 million years, which is very short compared with the age of the earth. Therefore the meteorites must be coeval with the solar system, and the plutonium must have been made close to, in both time and space, the event that gave birth to the sun.

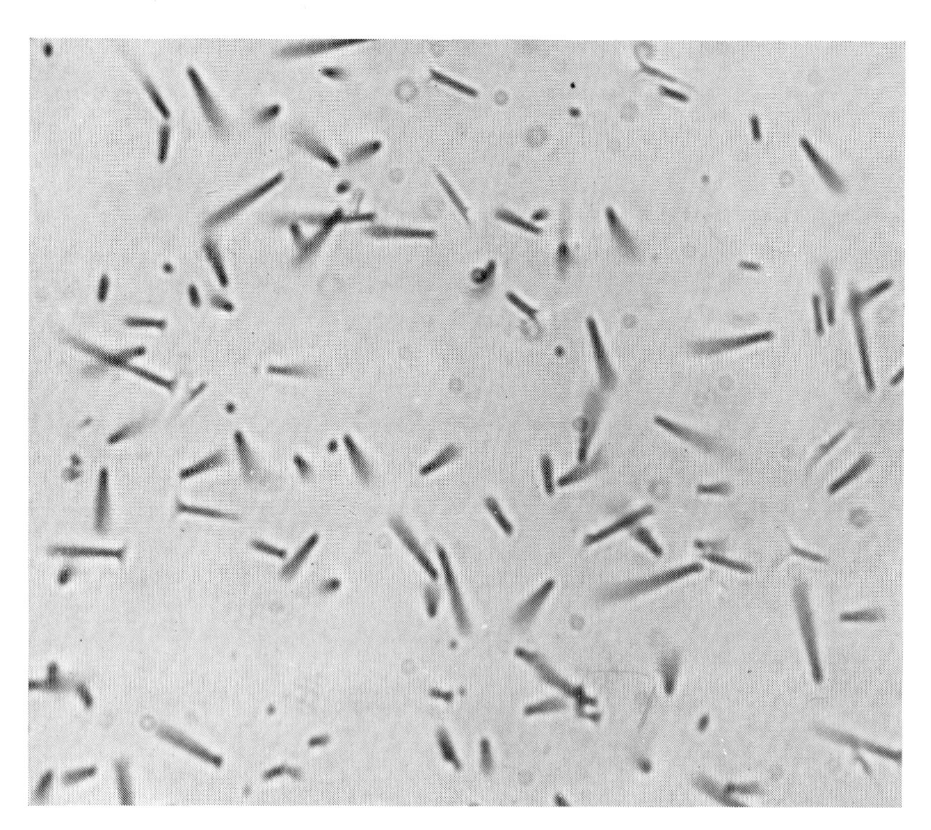

ANCIENT EVIDENCE that a locally violent environment existed in our galaxy immediately before the birth of the solar system is provided by photographs such as this one, which was made by P. B. Price of the University of California at Berkeley. The photograph shows radiation damage in the form of fission-fragment tracks made visible by etching in a crystal from a very ancient meteorite. Meteorites of this type do not contain enough uranium or thorium to account for the fission tracks. Instead the tracks appear to be the products of the spontaneous fission of the nucleus plutonium 244, which has a half-life of only 80 million years, a period that is very short compared with the age of the earth. Therefore the meteorites must be coeval with the solar system, and the plutonium must have been made close to, in both time and space, the event that gave birth to the solar system.

We are only beginning to understand the way stars and planets are born. It seems that stars are born in clusters of a few hundred or a few thousand at a time rather than singly. There is perhaps a cyclical rhythm in the life of a galaxy. For 100 million years the stars and the interstellar gas in any particular sector of a galaxy lie quiet. Then some kind of shock or gravitational wave passes by, compressing the gas and triggering gravitational condensation. Various hangups are overcome, and a large mass of gas condenses into new stars in a limited region of space. The most massive stars shine brilliantly for a few million years and die spectacularly as supernovas. The brief blaze of the clusters of short-lived massive stars makes the shock wave visible, from a distance of millions of light-years, as a bright spiral arm sweeping around the galaxy. After the massive new stars are burned out the less massive stars continue to condense, partially contaminated with plutonium. These more modest stars continue their quiet and frugal existence for billions of years after the spiral arm that gave them birth has passed by. In some such rhythm as this, 4.5 billion years ago, our solar system came into being.

Whether some similar rhythms, on an even more gigantic scale, are involved in the birth of the radio galaxies, the quasars and the nuclei of Seyfert galaxies we simply do not know. Each of these objects pours out quantities of energy millions of times greater than the output of the brightest supernova. We know nothing of their origins, and we know nothing of their effects on their surroundings. It would be strange if their effects did not ultimately turn out to be of major importance, both for science and for the history of life in the universe.

The main sources of energy available to us on the earth are chemical fuels, uranium and sunlight. In addition we hope one day to learn how to burn in a controlled fashion the deuterium in the oceans. All these energy stores exist here by virtue of hangups that have temporarily halted the universal processes of energy degradation. Sunlight is sustained by the thermonuclear, the weak-interaction and the opacity hangups. Uranium is preserved by the surface-tension hangup. Coal and oil have been buried in the ground and saved from oxidation by various biological and chemical hangups, the details of which are still under debate. Deuterium has been preserved in low abundance, after almost all of it was burned to form helium in the earliest stages of the history of the universe, because no thermonuclear reaction ever runs quite to completion.

Humanity is fortunate in having such a variety of energy resources at its disposal. In the very long run we shall need energy that is absolutely pollution-free; we shall have sunlight. In the fairly long run we shall need energy that is inexhaustible and moderately clean; we shall have deuterium. In the short run we shall need energy that is readily usable and abundant; we shall have uranium. Right now we need energy that is cheap and convenient; we have coal and oil. Nature has been kinder to us than we had any right to expect. As we look out into the universe and identify the many accidents of physics and astronomy that have worked together to our benefit, it almost seems as if the universe must in some sense have known that we were coming.

Since the Apollo voyages gave us a closeup view of the desolate landscape of the moon, many people have formed an impression of the earth as a uniquely beautiful and fragile oasis in a harsh and hostile universe. The distant pictures of the blue planet conveyed this impression most movingly. I wish to assert the contrary view. I believe the universe is friendly. I see no reason to suppose that the cosmic accidents that provided so abundantly for our welfare here on the earth will not do the same for us wherever else in the universe we choose to go.

Ko Fung was one of the great natural philosophers of ancient China. In the fourth century he wrote: "As for belief, there are things that are as clear as the sky, yet men prefer to sit under an upturned barrel." Some of the current discussions of the resources of mankind on the earth have a claustrophobic quality that Ko Fung's words describe very accurately. I hope that with this article I may have persuaded a few people to come out from under the barrel, and to look to the sky with hopeful eyes. I began with a quotation from Blake. Let me end with another from him, this time echoing the thought of Ko Fung: "If the doors of perception were cleansed every thing would appear to man as it is, infinite. For man has closed himself up, till he sees all things thro' narrow chinks of his cavern."

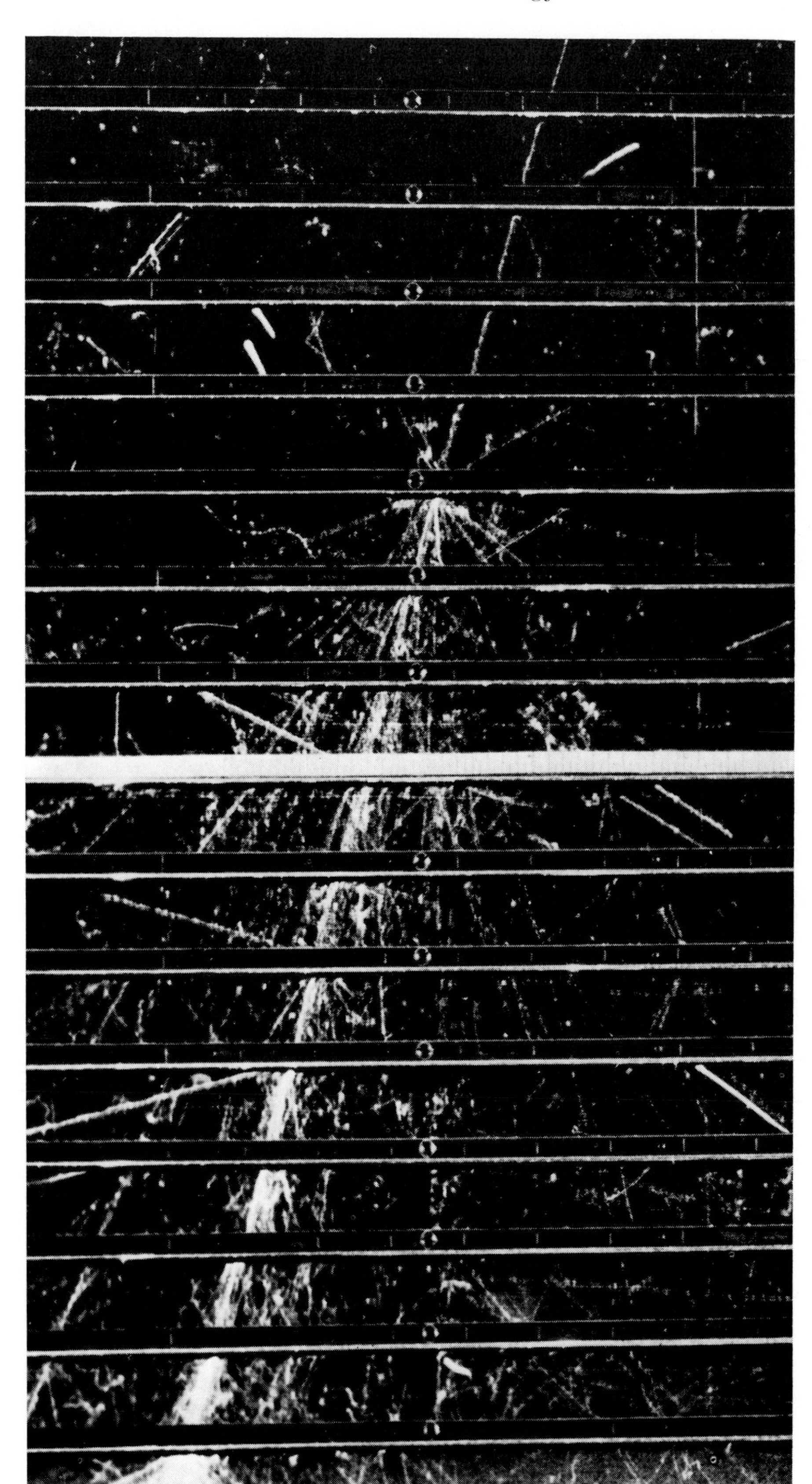

RECENT EVIDENCE of violent events in parts of the universe other than our own is contained in this cloud-chamber photograph of a primary cosmic ray track, obtained at an altitude of 17,200 feet on Mount Chacaltaya in Bolivia by Alfred Z. Hendel and his colleagues at the University of Michigan. The cloud chamber contained 17 iron plates, each half an inch thick. The cosmic ray, in this case a high-energy proton, entered the chamber from the top and passed through five plates before colliding with an iron nucleus in the sixth plate, producing a shower of secondary reaction products, mainly pi mesons. The energy of the incoming proton, approximately 1,100 billion electron volts, was measured by means of a detector mounted below the cloud chamber. Although it is not understood in detail where cosmic rays come from, they are known to carry about as much energy as starlight.

3

The Energy Resources of the Earth

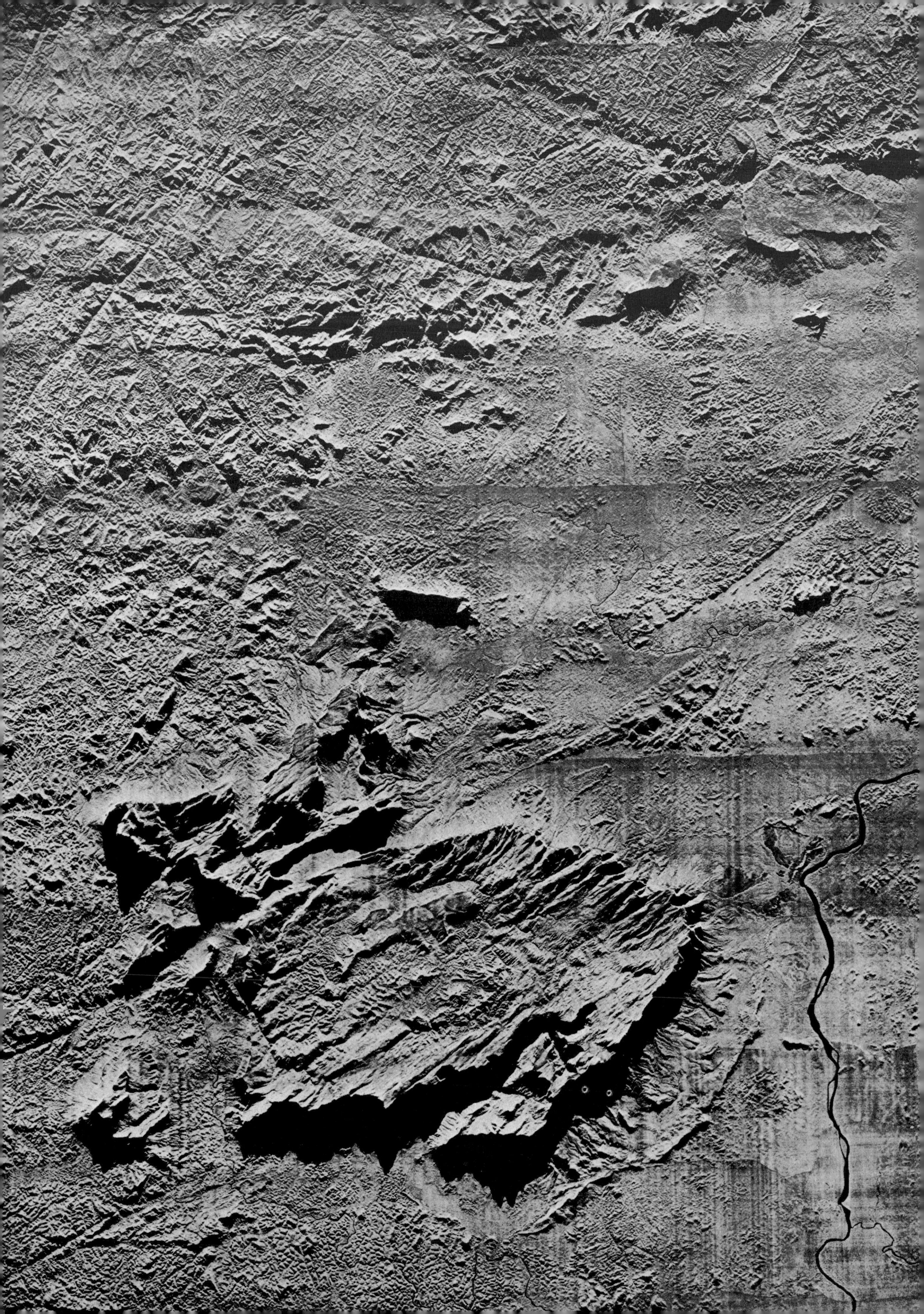

The Energy Resources of the Earth

M. KING HUBBERT

They are solar energy (current and stored), the tides, the earth's heat, fission fuels and possibly fusion fuels. From the standpoint of human history the epoch of the fossil fuels will be quite brief

Energy flows constantly into and out of the earth's surface environment. As a result the material constituents of the earth's surface are in a state of continuous or intermittent circulation. The source of the energy is preponderantly solar radiation, supplemented by small amounts of heat from the earth's interior and of tidal energy from the gravitational system of the earth, the moon and the sun. The materials of the earth's surface consist of the 92 naturally occurring chemical elements, all but a few of which behave in accordance with the principles of the conservation of matter and of nontransmutability as formulated in classical chemistry. A few of the elements or their isotopes, with abundances of only a few parts per million, are an exception to these principles in being radioactive. The exception is crucial in that it is the key to an additional large source of energy.

A small part of the matter at the earth's surface is embodied in living organisms: plants and animals. The leaves of the plants capture a small fraction of the incident solar radiation and store it chemically by the mechanism of photosynthesis. This store becomes the energy supply essential for the existence of the plant and animal kingdoms. Biologically stored energy is released by oxidation at a rate approximately equal to the rate of storage. Over millions of years, however, a minute fraction of the vegetable and animal matter is buried under conditions of incomplete oxidation and decay, thereby giving rise to the fossil fuels that provide most of the energy for industrialized societies.

It is difficult for people living now, who have become accustomed to the steady exponential growth in the consumption of energy from the fossil fuels, to realize how transitory the fossil-fuel epoch will eventually prove to be when it is viewed over a longer span of human history. The situation can better be seen in the perspective of some 10,000 years, half before the present and half afterward. On such a scale the complete cycle of the exploitation of the world's fossil fuels will be seen to encompass perhaps 1,300 years, with the principal segment of the cycle (defined as the period during which all but the first 10 percent and the last 10 percent of the fuels are extracted and burned) covering only about 300 years.

What, then, will provide industrial energy in the future on a scale at least as large as the present one? The answer lies in man's growing ability to exploit other sources of energy, chiefly nuclear at present but perhaps eventually the much larger source of solar energy. With this ability the energy resources now at hand are sufficient to sustain an industrial operation of the present magnitude for another millennium or longer. Moreover, with such resources of energy the limits to the growth of industrial activity are no longer set by a scarcity of energy but rather by the space and material limitations of a finite earth together with the principles of ecology. According to these principles both biological and industrial activities tend to increase exponentially with time, but the resources of the entire earth are not sufficient to sustain such an increase of any single component for more than a few tens of successive doublings.

Let us consider in greater detail the flow of energy through the earth's surface environment [*see illustration on next two pages*]. The inward flow of energy has three main sources: (1) the intercepted solar radiation; (2) thermal energy, which is conveyed to the surface of the earth from the warmer interior by the conduction of heat and by convection in hot springs and volcanoes, and (3) tidal energy, derived from the combined kinetic and potential energy of the earth-moon-sun system. It is possible in various ways to estimate approximately how large the input is from each source.

In the case of solar radiation the influx is expressed in terms of the solar constant, which is defined as the mean rate of flow of solar energy across a unit of area that is perpendicular to the radiation and outside the earth's atmosphere at the mean distance of the earth from the sun. Measurements made on the earth and in spacecraft give a mean value for the solar constant of 1.395 kilowatts per square meter, with a variation of about 2 percent. The total solar radiation intercepted by the earth's diametric plane of 1.275×10^{14} square meters is therefore 1.73×10^{17} watts.

The influx of heat by conduction from the earth's interior has been determined from measurements of the geothermal gradient (the increase of temperature with depth) and the thermal conductivity of the rocks involved. From thousands of such measurements, both on land and on the ocean beds, the average rate of flow of heat from the interior of the earth has been found to be about .063 watt per square meter. For the earth's surface area of 510×10^{12} square meters the total heat flow amounts to

RESOURCE EXPLORATION is beginning to be aided by airborne side-looking radar pictures such as the one on the opposite page made by the Aero Service Corporation and the Goodyear Aerospace Corporation. The technique has advantage of "seeing" through cloud cover and vegetation. This picture, which was made in southern Venezuela, extends 70 miles from left to right.

some 32×10^{12} watts. The rate of heat convection by hot springs and volcanoes is estimated to be only about 1 percent of the rate of conduction, or about $.3 \times 10^{12}$ watts.

The energy from tidal sources has been estimated at 3×10^{12} watts. When all three sources of energy are expressed in the common unit of 10^{12} watts, the total power influx into the earth's surface environment is found to be $173,035 \times 10^{12}$ watts. Solar radiation accounts for 99.98 percent of it. Another way of stating the sun's contribution to the energy budget of the earth is to note that at $173,000 \times 10^{12}$ watts it amounts to 5,000 times the energy input from all other sources combined.

About 30 percent of the incident solar energy ($52,000 \times 10^{12}$ watts) is directly reflected and scattered back into space as short-wavelength radiation. Another 47 percent ($81,000 \times 10^{12}$ watts) is absorbed by the atmosphere, the land surface and the oceans and converted directly into heat at the ambient surface temperature. Another 23 percent ($40,000 \times 10^{12}$ watts) is consumed in the evaporation, convection, precipitation and surface runoff of water in the hydrologic cycle. A small fraction, about 370×10^{12} watts, drives the atmospheric and oceanic convections and circulations and the ocean waves and is eventually dissipated into heat by friction. Finally, an even smaller fraction—about 40×10^{12} watts—is captured by the chlorophyll of plant leaves, where it becomes the essential energy supply of the photosynthetic process and eventually of the plant and animal kingdoms.

Photosynthesis fixes carbon in the leaf and stores solar energy in the form of carbohydrate. It also liberates oxygen and, with the decay or consumption of the leaf, dissipates energy. At any given time, averaged over a year or more, the balance between these processes is almost perfect. A minute fraction of the organic matter produced, however, is deposited in peat bogs or other oxygen-deficient environments under conditions that prevent complete decay and loss of energy.

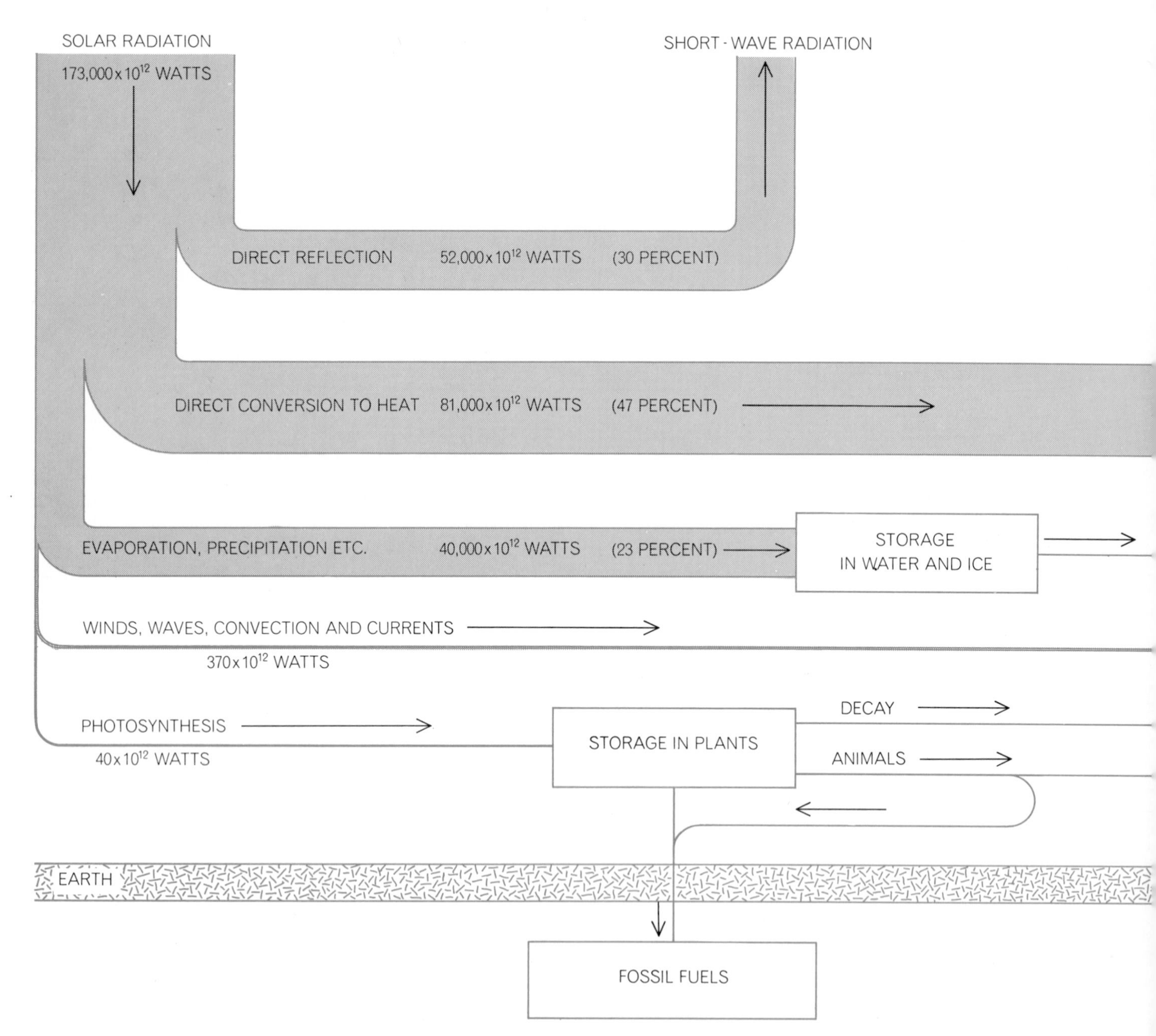

FLOW OF ENERGY to and from the earth is depicted by means of bands and lines that suggest by their width the contribution of each item to the earth's energy budget. The principal inputs are solar radiation, tidal energy and the energy from nuclear, thermal and gravitational sources. More than 99 percent of the input is solar radiation. The apportionment of incoming solar radiation is

Little of the organic material produced before the Cambrian period, which began about 600 million years ago, has been preserved. During the past 600 million years, however, some of the organic materials that did not immediately decay have been buried under a great thickness of sedimentary sands, muds and limes. These are the fossil fuels: coal, oil shale, petroleum and natural gas, which are rich in energy stored up chemically from the sunshine of the past 600 million years. The process is still continuing, but probably at about the same rate as in the past; the accumulation during the next million years will probably be a six-hundredth of the amount built up thus far.

Industrialization has of course withdrawn the deposits in this energy bank with increasing rapidity. In the case of coal, for example, the world's consumption during the past 110 years has been about 19 times greater than it was during the preceding seven centuries. The increasing magnitude of the rate of withdrawal can also be seen in the fact that the amount of coal produced and consumed since 1940 is approximately equal to the total consumption up to that time. The cumulative production from 1860 through 1970 was about 133 billion metric tons. The amount produced before 1860 was about seven million metric tons.

Petroleum and related products were not extracted in significant amounts before 1880. Since then production has increased at a nearly constant exponential rate. During the 80-year period from 1890 through 1970 the average rate of increase has been 6.94 percent per year, with a doubling period of 10 years. The cumulative production until the end of 1969 amounted to 227 billion (227×10^9) barrels, or 9.5 trillion U.S. gallons. Once again the period that encompasses most of the production is notably brief. The 102 years from 1857 to 1959 were required to produce the first half of the cumulative production; only the 10-year period from 1959 to 1969 was required for the second half.

Examining the relative energy contributions of coal and crude oil by comparing the heats of combustion of the respective fuels (in units of 10^{12} kilowatt-hours), one finds that until after 1900 the contribution from oil was barely significant compared with the contribution from coal. Since 1900 the contribution from oil has risen much faster than that from coal. By 1968 oil represented about 60 percent of the total. If the energy from natural gas and natural-gas liquids had been included, the contribution from petroleum would have been about 70 percent. In the U.S. alone 73 percent of the total energy produced from fossil fuels in 1968 was from petroleum and 27 percent from coal.

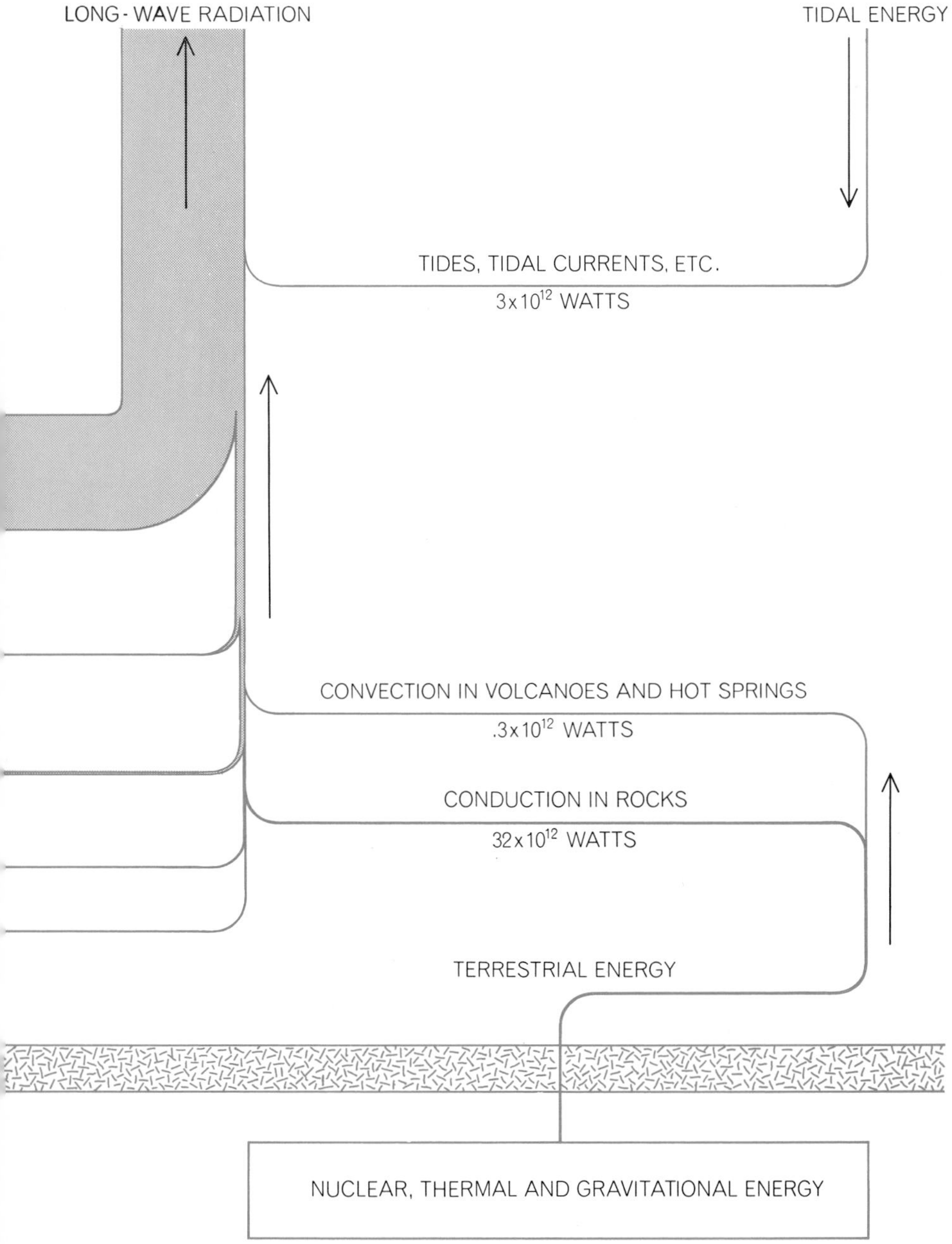

Broadly speaking, it can be said that the world's consumption of energy for industrial purposes is now doubling approximately once per decade. When confronted with a rate of growth of such magnitude, one can hardly fail to wonder how long it can be kept up. In the case of the fossil fuels a reasonably definite answer can be obtained. Their human exploitation consists of their being withdrawn from an essentially fixed initial supply. During their use as sources of energy they are destroyed. The complete cycle of exploitation of a fossil fuel must therefore have the following characteristics. Beginning at zero, the rate of production tends initially to increase exponentially. Then, as difficulties of discovery and extraction increase, the production rate slows in its growth, passes one maximum or more and, as the resource is progressively depleted, declines eventually to zero.

If known past and prospective future rates of production are combined with a reasonable estimate of the amount of a fuel initially present, one can calculate the probable length of time that the fuel can be exploited. In the case of coal reasonably good estimates of the

indicated by the horizontal bands beginning with "Direct reflection" and reading downward. The smallest portion goes to photosynthesis. Dead plants and animals buried in the earth give rise to fossil fuels, containing stored solar energy from millions of years past.

amount present in given regions can be made on the basis of geological mapping and a few widely spaced drill holes, inasmuch as coal is found in stratified beds or seams that are continuous over extensive areas. Such studies have been made in all the coal-bearing areas of the world.

The most recent compilation of the present information on the world's initial coal resources was made by Paul Averitt of the U.S. Geological Survey. His figures [*see illustration below*] represent minable coal, which is defined as 50 percent of the coal actually present. Included is coal in beds as thin as 14 inches (36 centimeters) and extending to depths of 4,000 feet (1.2 kilometers) or, in a few cases, 6,000 feet (1.8 kilometers).

Taking Averitt's estimate of an initial supply of 7.6 trillion metric tons and assuming that the present production rate of three billion metric tons per year does not double more than three times, one can expect that the peak in the rate of production will be reached sometime between 2100 and 2150. Disregarding the long time required to produce the first 10 percent and the last 10 percent, the length of time required to produce the middle 80 percent will be roughly the 300-year period from 2000 to 2300.

Estimating the amount of oil and gas that will ultimately be discovered and produced in a given area is considerably more hazardous than estimating for coal. The reason is that these fluids occur in restricted volumes of space and limited areas in sedimentary basins at all depths from a few hundred meters to more than eight kilometers. Nonetheless, the estimates for a given region improve as exploration and production proceed. In addition it is possible to make rough estimates for relatively undeveloped areas on the basis of geological comparisons between them and well-developed regions.

The most highly developed oil-producing region in the world is the coterminous area of the U.S.: the 48 states exclusive of Alaska and Hawaii. This area has until now led the world in petroleum development, and the U.S. is still the leading producer. For this region a large mass of data has been accumulated and a number of different methods of analysis have been developed that give fairly consistent estimates of the degree of advancement of petroleum exploration and of the amounts of oil and gas that may eventually be produced.

One such method is based on the principle that only a finite number of oil or gas fields existed initially in a given region. As exploration proceeds the shallowest and most evident fields are usually discovered first and the deeper and more obscure ones later. With each discovery the number of undiscovered fields decreases by one. The undiscovered fields are also likely to be deeper, more widely spaced and better concealed. Hence the amount of exploratory activity required to discover a fixed quantity of oil or gas steadily increases or, conversely, the average amount of oil or gas discovered for a fixed amount of exploratory activity steadily decreases.

Most new fields are discovered by what the industry calls "new-field wildcat wells," meaning wells drilled in new territory that is not in the immediate vicinity of known fields. In the U.S. statistics have been kept annually since 1945 on the number of new-field wildcat wells required to make one significant discovery of oil or gas ("significant" being defined as one million barrels of oil or an equivalent amount of gas). The discoveries for a given year are evaluated only after six years of subsequent development. In 1945 it required 26

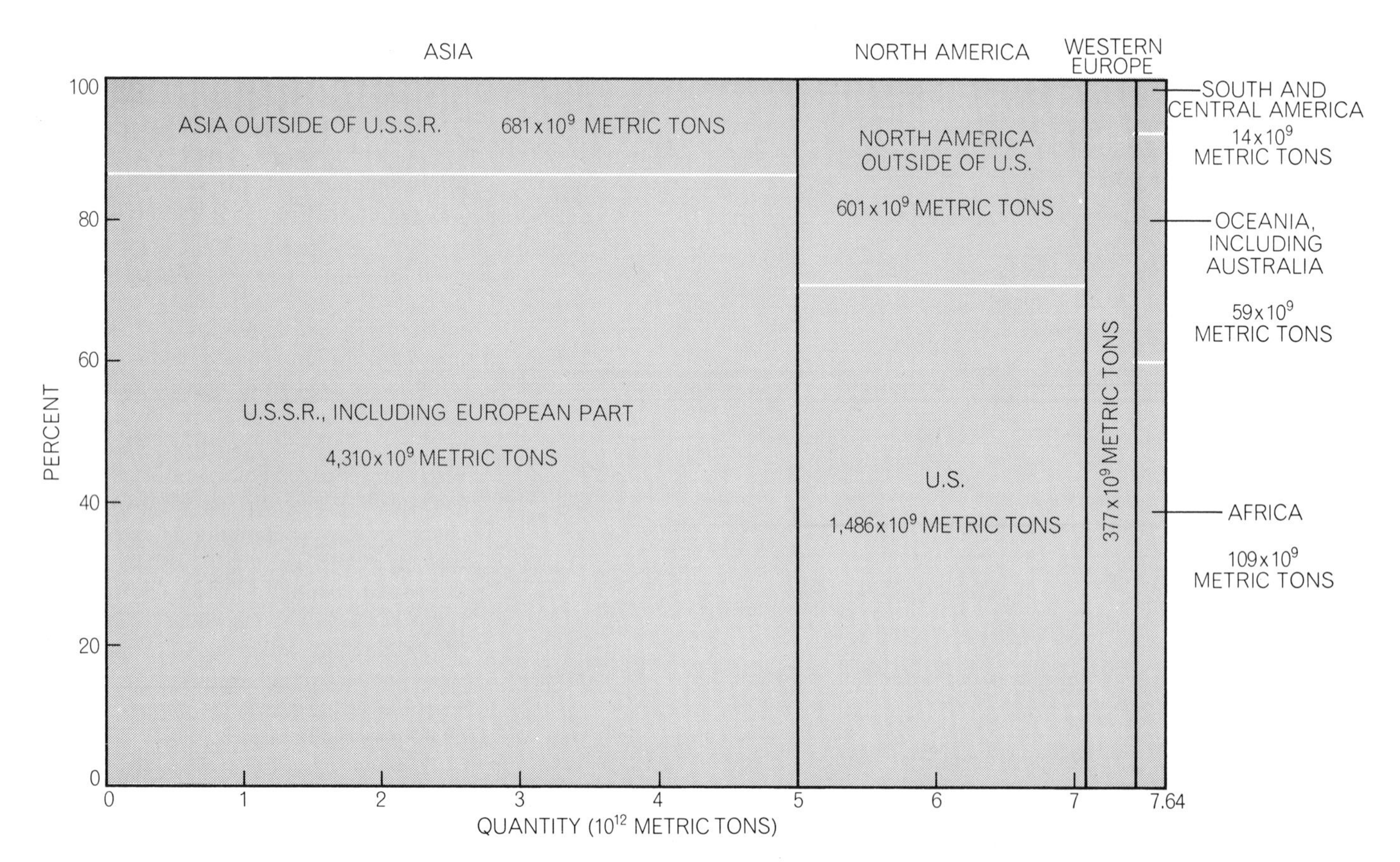

COAL RESOURCES of the world are indicated on the basis of data compiled by Paul Averitt of the U.S. Geological Survey. The figures represent the total initial resources of minable coal, which is defined as 50 percent of the coal actually present. The horizontal scale gives the total supply. Each vertical block shows the apportionment of the supply in a continent. From the first block, for example, one can ascertain that Asia has some 5×10^{12} metric tons of minable coal, of which about 86 percent is in the U.S.S.R.

new-field wildcat wells to make a significant discovery; by 1963 the number had increased to 65.

Another way of illuminating the problem is to consider the amount of oil discovered per foot of exploratory drilling. From 1860 to 1920, when oil was fairly easy to find, the ratio was 194 barrels per foot. From 1920 to 1928 the ratio declined to 167 barrels per foot. Between 1928 and 1938, partly because of the discovery of the large East Texas oil field and partly because of new exploratory techniques, the ratio rose to its maximum of 276 barrels per foot. Since then it has fallen sharply to a nearly constant rate of about 35 barrels per foot. Yet the period of this decline coincided with the time of the most intensive research and development in petroleum exploration and production in the history of the industry.

The cumulative discoveries in the 48 states up to 1965 amounted to 136 billion barrels. From this record of drilling and discovery it can be estimated that the ultimate total discoveries in the coterminous U.S. and the adjacent continental shelves will be about 165 billion barrels. The discoveries up to 1965 therefore represent about 82 percent of the prospective ultimate total. Making due allowance for the range of uncertainty in estimates of future discovery, it still appears that at least 75 percent of the ultimate amount of oil to be produced in this area will be obtained from fields that had already been discovered by 1965.

For natural gas in the 48 states the present rate of discovery, averaged over a decade, is about 6,500 cubic feet per barrel of oil. Assuming the same ratio for the estimated ultimate amount of 165 billion barrels of crude oil, the ultimate amount of natural gas would be about 1,075 trillion cubic feet. Combining the estimates for oil and gas with the trends of production makes it possible to estimate how long these energy resources will last. In the case of oil the period of peak production appears to be the present. The time span required to produce the middle 80 percent of the ultimate cumulative production is approximately the 65-year period from 1934 to 1999—less than the span of a human lifetime. For natural gas the peak of production will probably be reached between 1975 and 1980.

The discoveries of petroleum in Alaska modify the picture somewhat. In particular the field at Prudhoe Bay appears likely by present estimates to contain about 10 billion barrels, making it twice as large as the East Texas field, which was the largest in the U.S. previously. Only a rough estimate can be made of the eventual discoveries of petroleum in Alaska. Such a speculative estimate would be from 30 to 50 billion barrels. One must bear in mind, however, that 30 billion barrels is less than a 10-year supply for the U.S. at the present rate of consumption. Hence it appears likely that the principal effect of the oil from Alaska will be to retard the rate of decline of total U.S. production rather than to postpone the date of its peak.

Estimates of ultimate world production of oil range from 1,350 billion barrels to 2,100 billion barrels. For the higher figure the peak in the rate of world production would be reached about the year 2000. The period of consumption of the middle 80 percent will probably be some 58 to 64 years, depending on whether the lower or the higher estimate is used [*see bottom illustration on page 39*].

A substantial but still finite amount of oil can be extracted from tar sands and oil shales, where production has barely begun. The largest tar-sand deposits are in northern Alberta; they have total recoverable reserves of about 300 billion

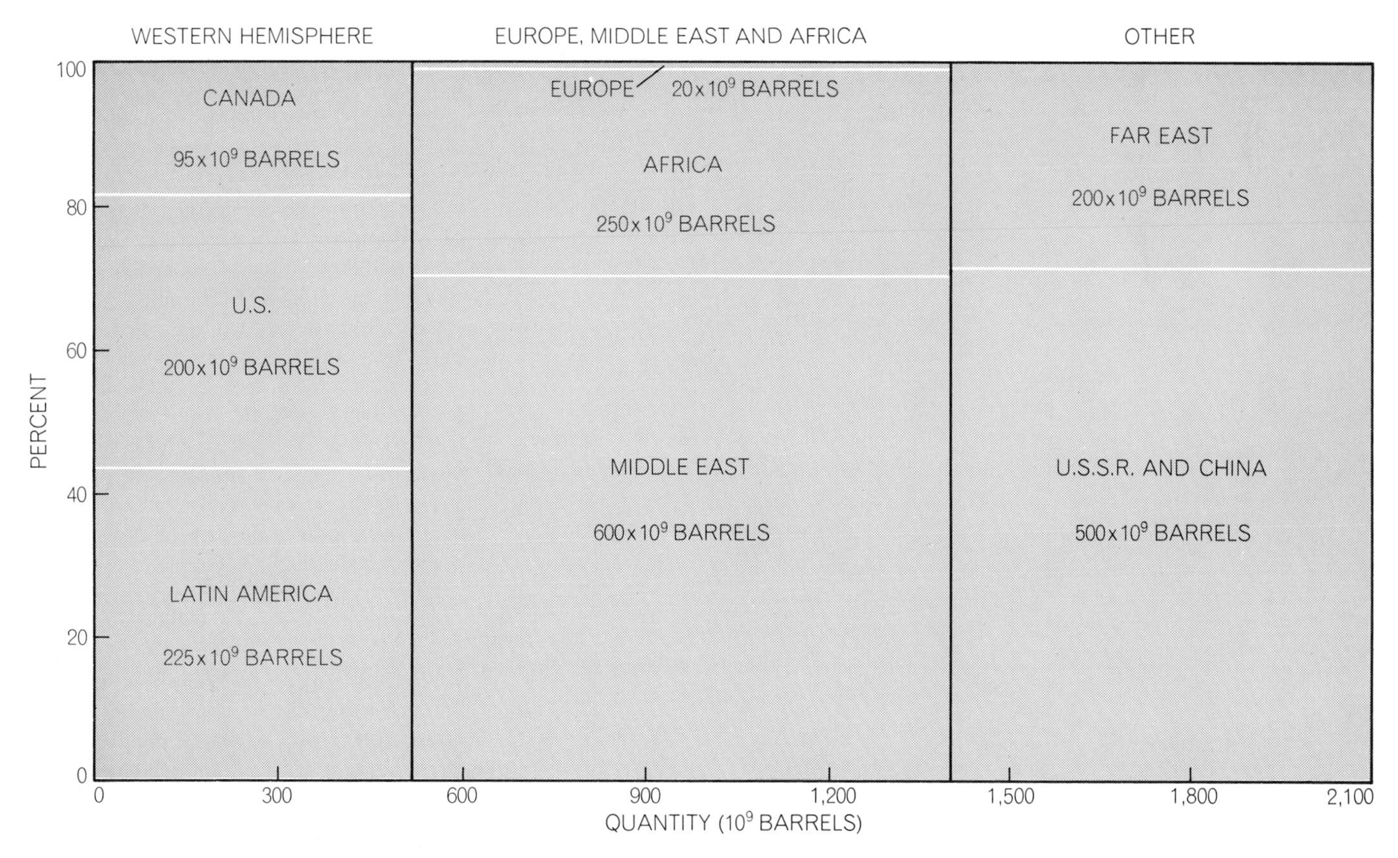

PETROLEUM RESOURCES of the world are depicted in an arrangement that can be read in the same way as the diagram of coal supplies on the opposite page. The figures for petroleum are derived from estimates made in 1967 by W. P. Ryman of the Standard Oil Company of New Jersey. They represent ultimate crude-oil production, including oil from offshore areas, and consist of oil already produced, proved and probable reserves, and future discoveries. Estimates as low as 1,350 $\times$ 10^9 barrels have also been made.

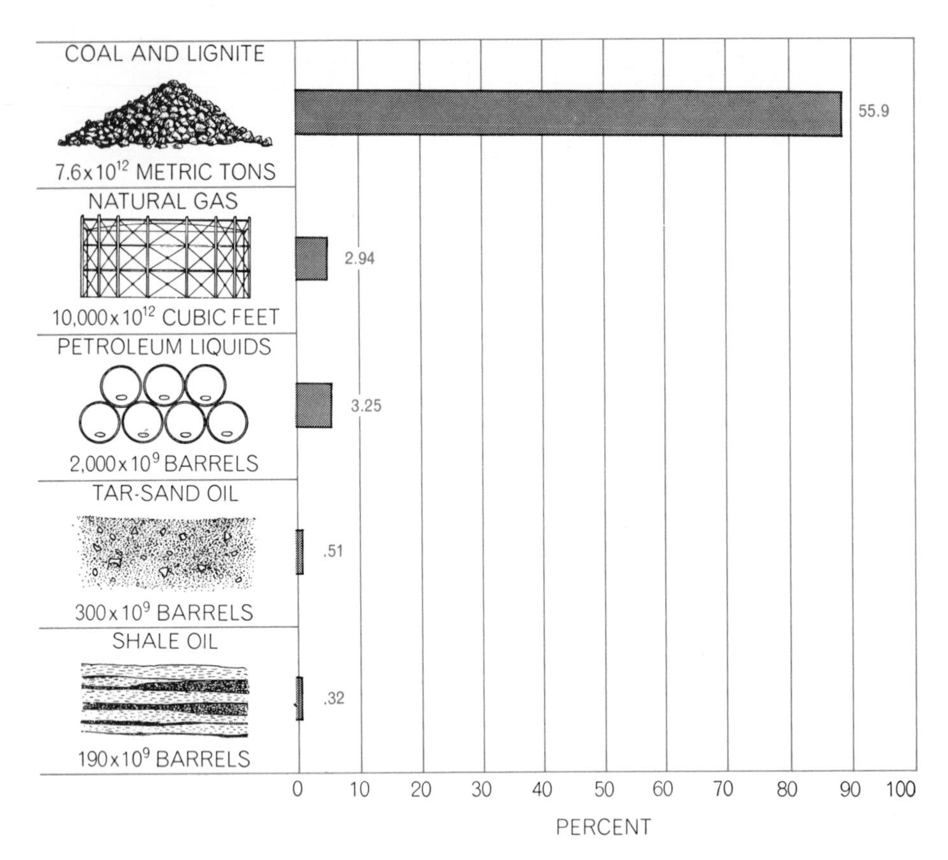

ENERGY CONTENT of the world's initial supply of recoverable fossil fuels is given in units of 10^{15} thermal kilowatt-hours (*color*). Coal and lignite, for example, contain 55.9 × 10^{15} thermal kilowatt-hours of energy and represent 88.8 percent of the recoverable energy.

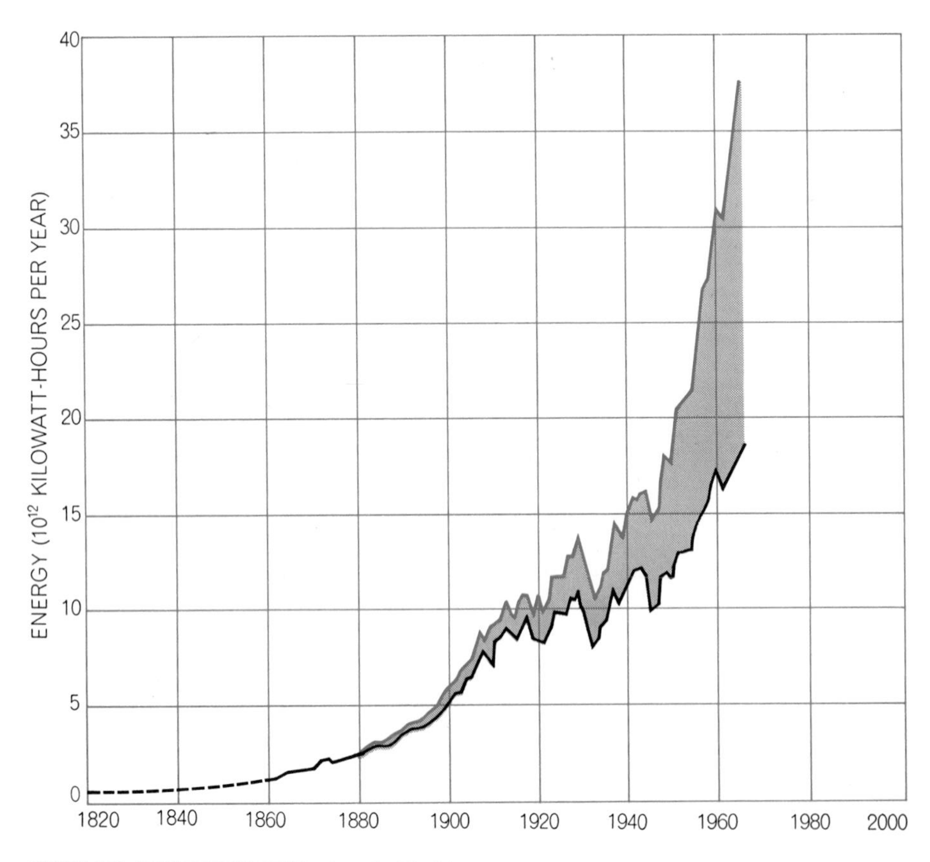

ENERGY CONTRIBUTION of coal (*black*) and coal plus oil (*color*) is portrayed in terms of their heat of combustion. Before 1900 the energy contribution from oil was barely significant. Since then the contribution from oil (*shaded area*) has risen much more rapidly than that from coal. By 1968 oil represented about 60 percent of the total. If the energy from natural gas were included, petroleum would account for about 70 percent of the total.

barrels. A world summary of oil shales by Donald C. Duncan and Vernon E. Swanson of the U.S. Geological Survey indicated a total of about 3,100 billion barrels in shales containing from 10 to 100 gallons per ton, of which 190 billion barrels were considered to be recoverable under 1965 conditions.

Since the fossil fuels will inevitably be exhausted, probably within a few centuries, the question arises of what other sources of energy can be tapped to supply the power requirements of a moderately industrialized world after the fossil fuels are gone. Five forms of energy appear to be possibilities: solar energy used directly, solar energy used indirectly, tidal energy, geothermal energy and nuclear energy.

Until now the direct use of solar power has been on a small scale for such purposes as heating water and generating electricity for spacecraft by means of photovoltaic cells. Much more substantial installations will be needed if solar power is to replace the fossil fuels on an industrial scale. The need would be for solar power plants in units of, say, 1,000 megawatts. Moreover, because solar radiation is intermittent at a fixed location on the earth, provision must also be made for large-scale storage of energy in order to smooth out the daily variation.

The most favorable sites for developing solar power are desert areas not more than 35 degrees north or south of the Equator. Such areas are to be found in the southwestern U.S., the region extending from the Sahara across the Arabian Peninsula to the Persian Gulf, the Atacama Desert in northern Chile and central Australia. These areas receive some 3,000 to 4,000 hours of sunshine per year, and the amount of solar energy incident on a horizontal surface ranges from 300 to 650 calories per square centimeter per day. (Three hundred calories, the winter minimum, amounts when averaged over 24 hours to a mean power density of 145 watts per square meter.)

Three schemes for collecting and converting this energy in a 1,000-megawatt plant can be considered. The first involves the use of flat plates of photovoltaic cells having an efficiency of about 10 percent. A second possibility is a recent proposal by Aden B. Meinel and Marjorie P. Meinel of the University of Arizona for utilizing the hothouse effect by means of selective coatings on pipes carrying a molten mixture of sodium and potassium raised by solar energy to a temperature of 540 degrees Celsius. By

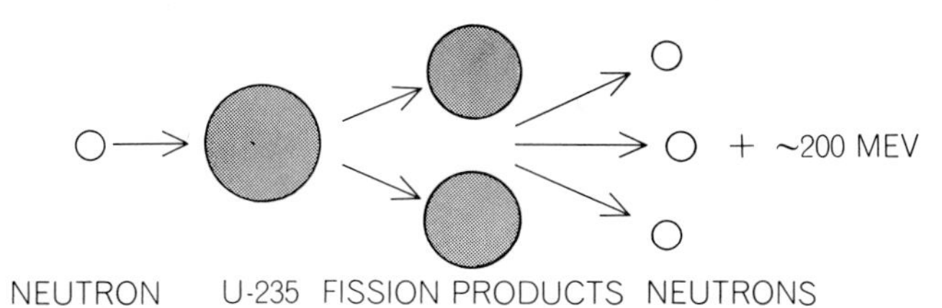

FISSION AND FUSION REACTIONS hold the promise of serving as sources of energy when fossil fuels are depleted. Present nuclear-power plants burn uranium 235 as a fuel. Breeder reactors now under development will be able to use surplus neutrons from the fission of uranium 235 (*left*) to create other nuclear fuels: plutonium 239 and uranium 233. Two promising fusion reactions, deuterium-deuterium and deuterium-tritium, are at right. The energy released by the various reactions is shown in million electron volts.

means of a heat exchanger this heat is stored at a constant temperature in an insulated chamber filled with a mixture of sodium and potassium chlorides that has enough heat capacity for at least one day's collection. Heat extracted from this chamber operates a conventional steam-electric power plant. The computed efficiency for this proposal is said to be about 30 percent.

A third system has been proposed by Alvin F. Hildebrandt and Gregory M. Haas of the University of Houston. It entails reflecting the radiation reaching a square-mile area into a solar furnace and boiler at the top of a 1,500-foot tower. Heat from the boiler at a temperature of 2,000 degrees Kelvin would be converted into electric power by a magnetohydrodynamic conversion. An energy-storage system based on the hydrolysis of water is also proposed. An overall efficiency of about 20 percent is estimated.

Over the range of efficiencies from 10 to 30 percent the amount of thermal power that would have to be collected for a 1,000-megawatt plant would range from 10,000 to 3,300 thermal megawatts. Accordingly the collecting areas for the three schemes would be 70, 35 and 23 square kilometers respectively. With the least of the three efficiencies the area required for an electric-power capacity of 350,000 megawatts—the approximate capacity of the U.S. in 1970—would be 24,500 square kilometers, which is somewhat less than a tenth of the area of Arizona.

The physical knowledge and technological resources needed to use solar energy on such a scale are now available. The technological difficulties of doing so, however, should not be minimized.

Using solar power indirectly means relying on the wind, which appears impractical on a large scale, or on the streamflow part of the hydrologic cycle. At first glance the use of streamflow appears promising, because the world's total water-power capacity in suitable sites is about three trillion watts, which approximates the present use of energy in industry. Only 8.5 percent of the water power is developed at present, however, and the three regions with the greatest potential—Africa, South America and Southeast Asia—are the least developed industrially. Economic problems therefore stand in the way of extensive development of additional water power.

Tidal power is obtained from the filling and emptying of a bay or an estuary that can be closed by a dam. The enclosed basin is allowed to fill and empty only during brief periods at high and low tides in order to develop as much power as possible. A number of promising sites exist; their potential capacities range from two megawatts to 20,000 megawatts each. The total potential tidal power, however, amounts to about 64 billion watts, which is only 2 percent of the world's potential water power. Only one full-scale tidal-electric plant has been built; it is on the Rance estuary on the Channel Island coast of France. Its capacity at start-up in 1966 was 240 megawatts; an ultimate capacity of 320 megawatts is planned.

Geothermal power is obtained by extracting heat that is temporarily stored in the earth by such sources as volcanoes and the hot water filling the sands of deep sedimentary basins. Only volcanic sources are significantly exploited at present. A geothermal-power operation has been under way in the Larderello area of Italy since 1904 and now has a capacity of 370 megawatts. The two other main areas of geothermal-power production are The Geysers in northern California and Wairakei in New Zealand. Production at The Geysers began in 1960 with a 12.5-megawatt unit. By 1969 the capacity had reached 82 megawatts, and plans are to reach a total installed capacity of 400 megawatts by 1973. The Wairakei plant began operation in 1958 and now has a capacity of 290 megawatts, which is believed to be about the maximum for the site.

Donald E. White of the U.S. Geological Survey has estimated that the stored thermal energy in the world's major geothermal areas amounts to about 4×10^{20} joules. With a 25 percent conversion factor the production of electrical energy would be about 10^{20} joules, or three million megawatt-years. If this energy, which is depletable, were withdrawn over a period of 50 years, the average annual power production would be 60,000 megawatts, which is comparable to the potential tidal power.

Nuclear power must be considered under the two headings of fission and fusion. Fission involves the splitting of nuclei of heavy elements such as uranium. Fusion involves the combining of light nuclei such as deuterium. Uranium 235, which is a rare isotope (each 100,000 atoms of natural uranium include six atoms of uranium 234, 711 atoms of uranium 235 and 99,283 atoms of uranium 238), is the only atomic species capable of fissioning under relatively mild environmental conditions. If nuclear energy depended entirely on uranium 235, the nuclear-fuel epoch would be brief. By breeding, however, wherein by absorbing neutrons in a nuclear reactor uranium 238 is transformed into fissionable plutonium 239 or thorium 232 becomes fissionable uranium 233, it is possible to create more nuclear fuel than is consumed. With breeding the entire supply of natural uranium and thorium would thus become available as fuel for fission reactors.

Most of the reactors now operating or planned in the rapidly growing nuclear-power industry in the U.S. depend essentially on uranium 235. The U.S. Atomic Energy Commission has estimated that the uranium requirement to meet the projected growth rate from 1970 to 1980 is 206,000 short tons of uranium oxide (U_3O_8). A report recently issued by the European Nuclear Energy Agency and the International Atomic Energy Agency projects requirements of 430,000 short tons of uranium

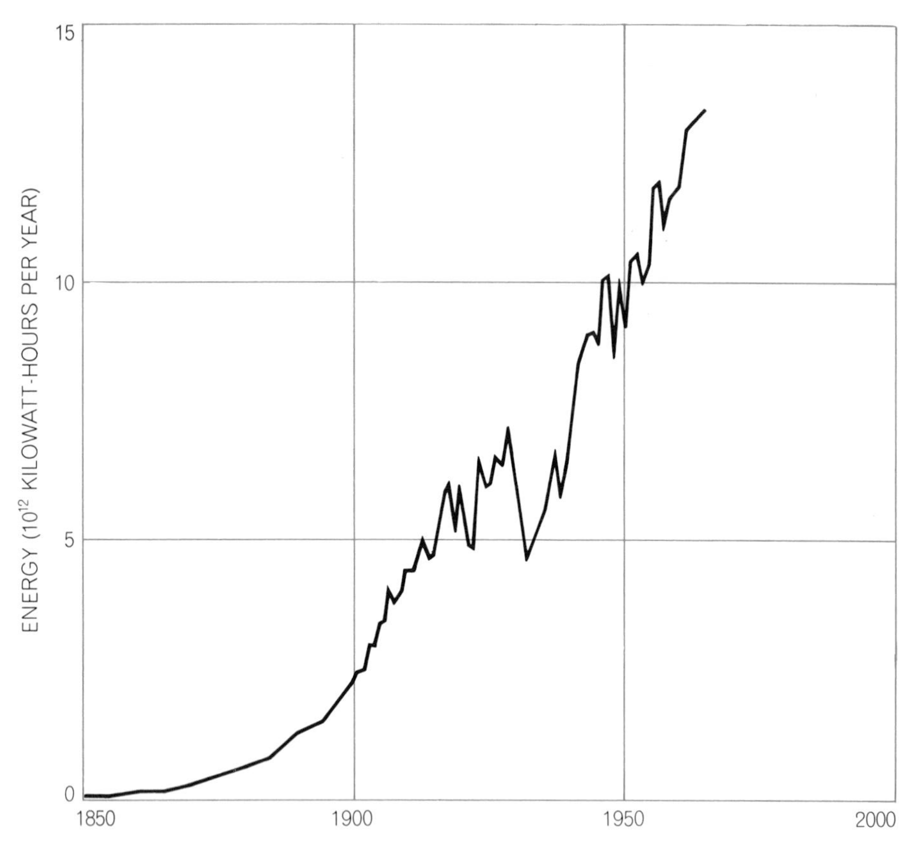

U.S. PRODUCTION OF ENERGY from coal, from petroleum and related sources, from water power and from nuclear reactors is charted for 120 years. The petroleum increment includes natural gas and associated liquids. The dip at center reflects impact of Depression.

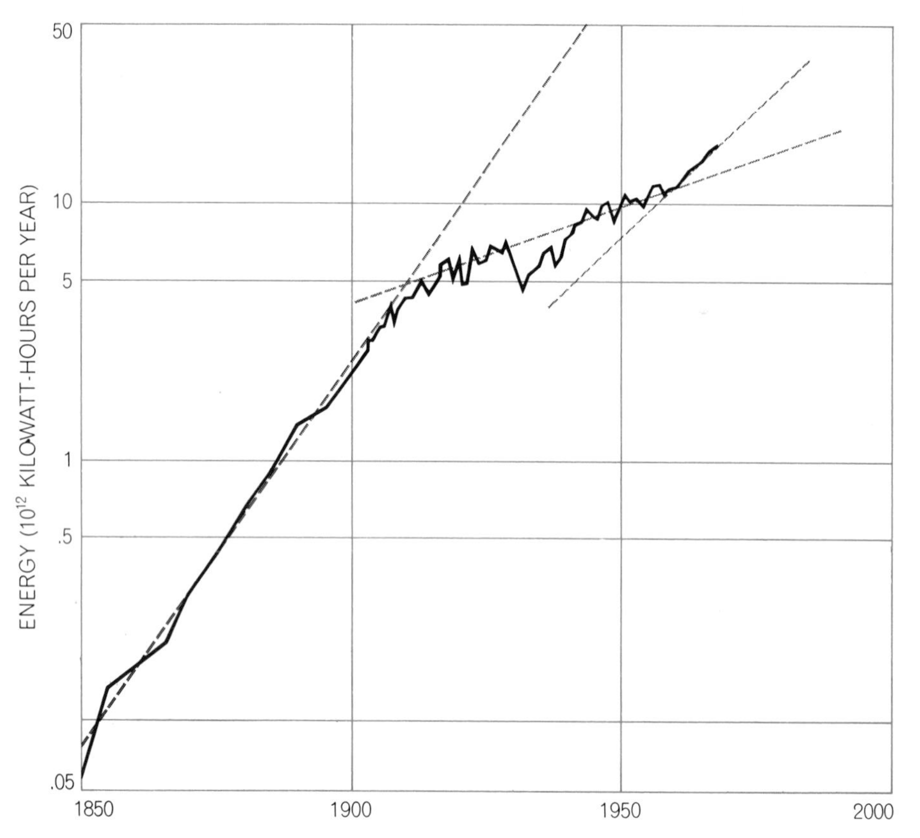

RATE OF GROWTH of U.S. energy production is shown by plotting on a semilogarithmic scale the data represented in the illustration at the top of the page. Broken lines show that the rise had three distinct periods. In the first the growth rate was 6.91 percent per year and the doubling period was 10 years; in the second the rate was 1.77 percent and the doubling period was 39 years; in the third the rate was 4.25 percent with doubling in 16.3 years.

oxide for the non-Communist nations during the same period.

Against these requirements the AEC estimates that the quantity of uranium oxide producible at $8 per pound from present reserves in the U.S. is 243,000 tons, and the world reserves at $10 per pound or less are estimated in the other report at 840,000 tons. The same report estimates that to meet future requirements additional reserves of more than a million short tons will have to be discovered and developed by 1985.

Although new discoveries of uranium will doubtless continue to be made (a large one was recently reported in northeastern Australia), all present evidence indicates that without a transition to breeder reactors an acute shortage of low-cost ores is likely to develop before the end of the century. Hence an intensive effort to develop large-scale breeder reactors for power production is in progress. If it succeeds, the situation with regard to fuel supply will be drastically altered.

This prospect results from the fact that with the breeder reactor the amount of energy obtainable from one gram of uranium 238 amounts to 8.1×10^{10} joules of heat. That is equal to the heat of combustion of 2.7 metric tons of coal or 13.7 barrels (1.9 metric tons) of crude oil. Disregarding the rather limited supplies of high-grade uranium ore that are available, let us consider the much more abundant low-grade ores. One example will indicate the possibilities.

The Chattanooga black shale (of Devonian age) crops out along the western edge of the Appalachian Mountains in eastern Tennessee and underlies at minable depths most of Tennessee, Kentucky, Ohio, Indiana and Illinois. In its outcrop area in eastern Tennessee this shale contains a layer about five meters thick that has a uranium content of about 60 grams per metric ton. That amount of uranium is equivalent to about 162 metric tons of bituminous coal or 822 barrels of crude oil. With the density of the rock some 2.5 metric tons per cubic meter, a vertical column of rock five meters long and one square meter in cross section would contain 12.5 tons of rock and 750 grams of uranium. The energy content of the shale per square meter of surface area would therefore be equivalent to about 2,000 tons of coal or 10,000 barrels of oil. Allowing for a 50 percent loss in mining and extracting the uranium, we are still left with the equivalent of 1,000 tons of coal or 5,000 barrels of oil per square meter.

Taking Averitt's estimate of 1.5 trillion metric tons for the initial minable coal in the U.S. and a round figure of 250 billion barrels for the petroleum liquids, we find that the nuclear energy in an area of about 1,500 square kilometers of Chattanooga shale would equal the energy in the initial minable coal; 50 square kilometers would hold the energy equivalent of the petroleum liquids. Adding natural gas and oil shales, an area of roughly 2,000 square kilometers of Chattanooga shale would be equivalent to the initial supply of all the fossil fuels in the U.S. The area is about 2 percent of the area of Tennessee and a very small fraction of the total area underlain by the shale. Many other low-grade deposits of comparable magnitude exist. Hence by means of the breeder reactor the energy potentially available from the fissioning of uranium and thorium is at least a few orders of magnitude greater than that from all the fossil fuels combined.

David J. Rose of the AEC, reviewing recently the prospects for controlled fusion, found the deuterium-tritium reaction to be the most promising. Deuterium is abundant (one atom to each 6,700 atoms of hydrogen), and the energy cost of separating it would be almost negligible compared with the amount of energy released by fusion. Tritium, on the other hand, exists only in tiny amounts in nature. Larger amounts must be made from lithium 6 and lithium 7 by nuclear bombardment. The limiting isotope is lithium 6, which has an abundance of only 7.4 percent of natural lithium.

Considering the amount of hydrogen in the oceans, deuterium can be regarded as superabundant. It can also be extracted easily. Lithium is much less

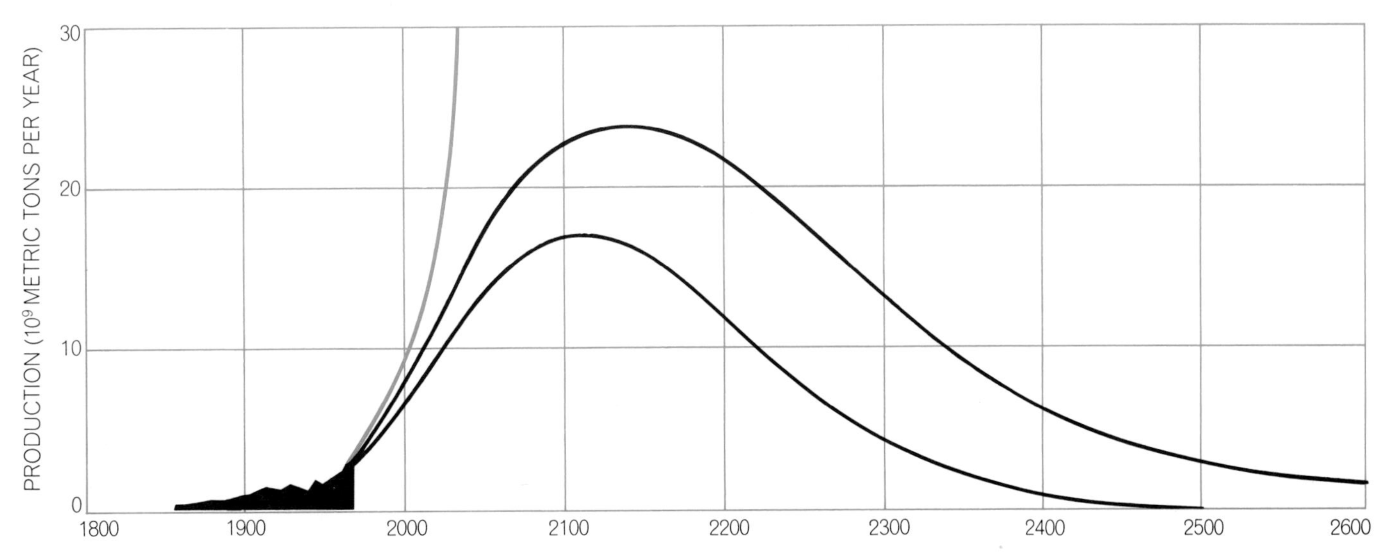

CYCLE OF WORLD COAL PRODUCTION is plotted on the basis of estimated supplies and rates of production. The top curve reflects Averitt's estimate of 7.6×10^{12} metric tons as the initial supply of minable coal; the bottom curve reflects an estimate of 4.3×10^{12} metric tons. The curve that rises to the top of the graph shows the trend if production continued to rise at the present rate of 3.56 percent per year. The amount of coal mined and burned in the century beginning in 1870 is shown by the black area at left.

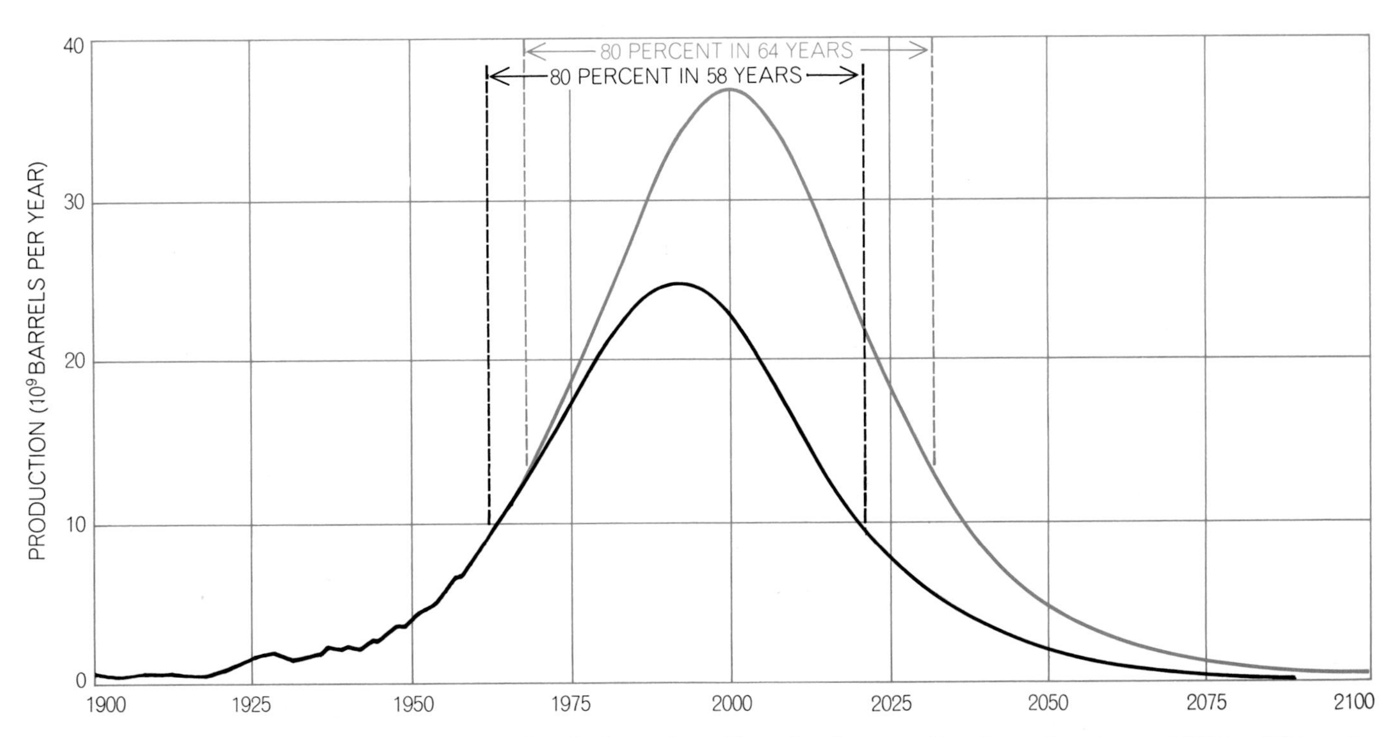

CYCLE OF WORLD OIL PRODUCTION is plotted on the basis of two estimates of the amount of oil that will ultimately be produced. The colored curve reflects Ryman's estimate of $2,100 \times 10^9$ barrels and the black curve represents an estimate of $1,350 \times 10^9$ barrels.

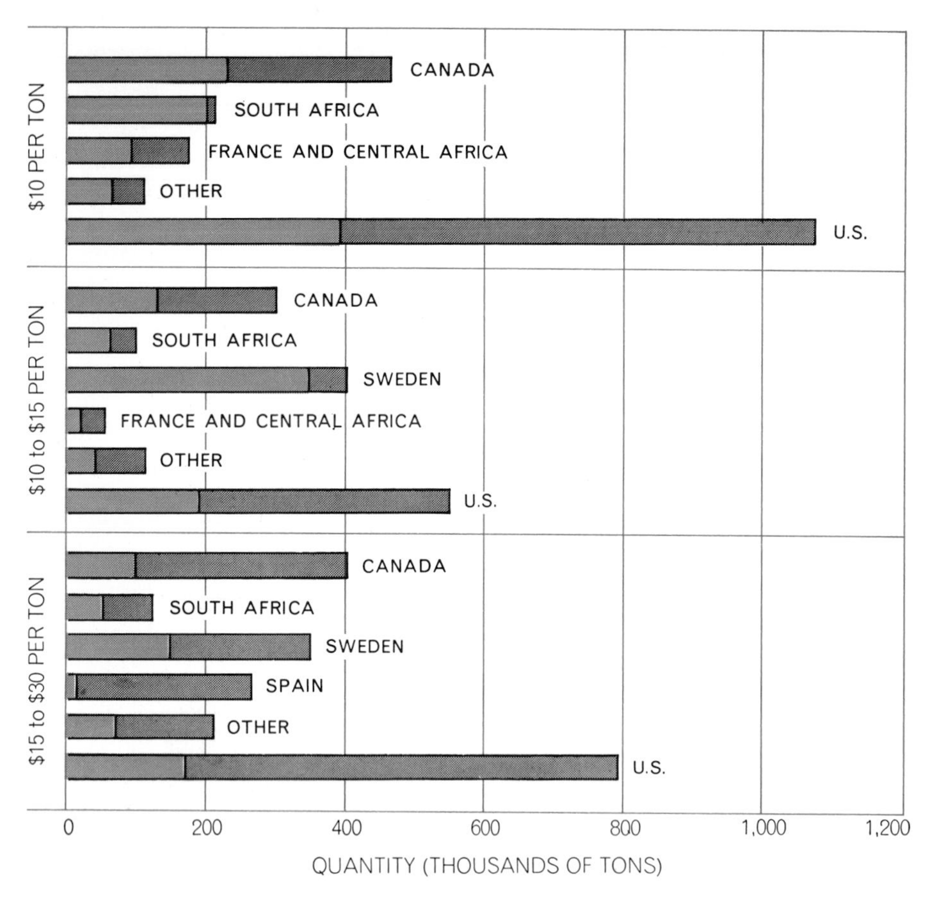

WORLD RESERVES OF URANIUM, which would be the source of nuclear power derived from atomic fission, are given in tons of uranium oxide (U_3O_8). The colored part of each bar represents reasonably assured supplies and the gray part estimated additional supplies.

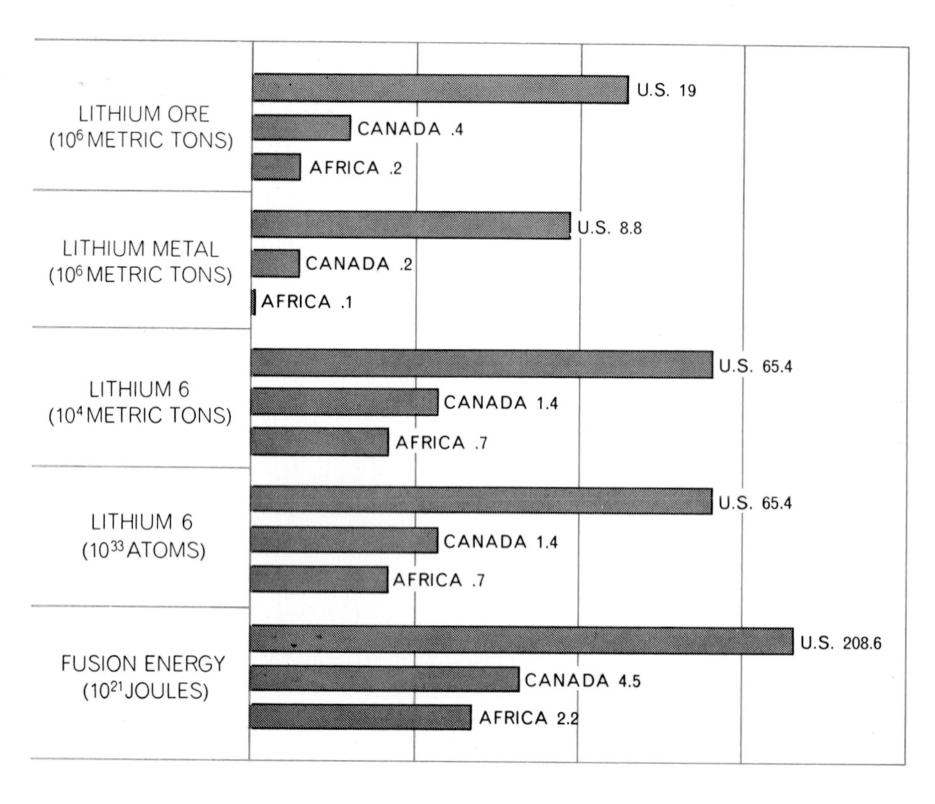

WORLD RESERVES OF LITHIUM, which would be the limiting factor in the deuterium-tritium fusion reaction, are stated in terms of lithium 6 because it is the least abundant isotope. Even with this limitation the energy obtainable from fusion through the deuterium-tritium reaction would almost equal the energy content of the world's fossil-fuel supply.

abundant. It is produced from the geologically rare igneous rocks known as pegmatites and from the salts of saline lakes. The measured, indicated and inferred lithium resources in the U.S., Canada and Africa total 9.1 million tons of elemental lithium, of which the content of lithium 6 would be 7.42 atom percent, or 67,500 metric tons. From this amount of lithium 6 the fusion energy obtainable at 3.19×10^{-12} joule per atom would be 215×10^{21} joules, which is approximately equal to the energy content of the world's fossil fuels.

As long as fusion power is dependent on the deuterium-tritium reaction, which at present appears to be somewhat the easier because it proceeds at a lower temperature, the energy obtainable from this source appears to be of about the same order of magnitude as that from fossil fuels. If fusion can be accomplished with the deuterium-deuterium reaction, the picture will be markedly changed. By this reaction the energy released per deuterium atom consumed is 7.94×10^{-13} joule. One cubic meter of water contains about 10^{25} atoms of deuterium having a mass of 34.4 grams and a potential fusion energy of 7.94×10^{12} joules. This is equivalent to the heat of combustion of 300 metric tons of coal or 1,500 barrels of crude oil. Since a cubic kilometer contains 10^9 cubic meters, the fuel equivalents of one cubic kilometer of seawater are 300 billion tons of coal or 1,500 billion barrels of crude oil. The total volume of the oceans is about 1.5 billion cubic kilometers. If enough deuterium were withdrawn to reduce the initial concentration by 1 percent, the energy released by fusion would amount to about 500,000 times the energy of the world's initial supply of fossil fuels!

Unlimited resources of energy, however, do not imply an unlimited number of power plants. It is as true of power plants or automobiles as it is of biological populations that the earth cannot sustain any physical growth for more than a few tens of successive doublings. Because of this impossibility the exponential rates of industrial and population growth that have prevailed during the past century and a half must soon cease. Although the forthcoming period of stability poses no insuperable physical or biological difficulties, it can hardly fail to force a major revision of those aspects of our current social and economic thinking that stem from the assumption that the growth rates that have characterized this temporary period can somehow be made permanent.

4

The Flow of Energy in the Biosphere

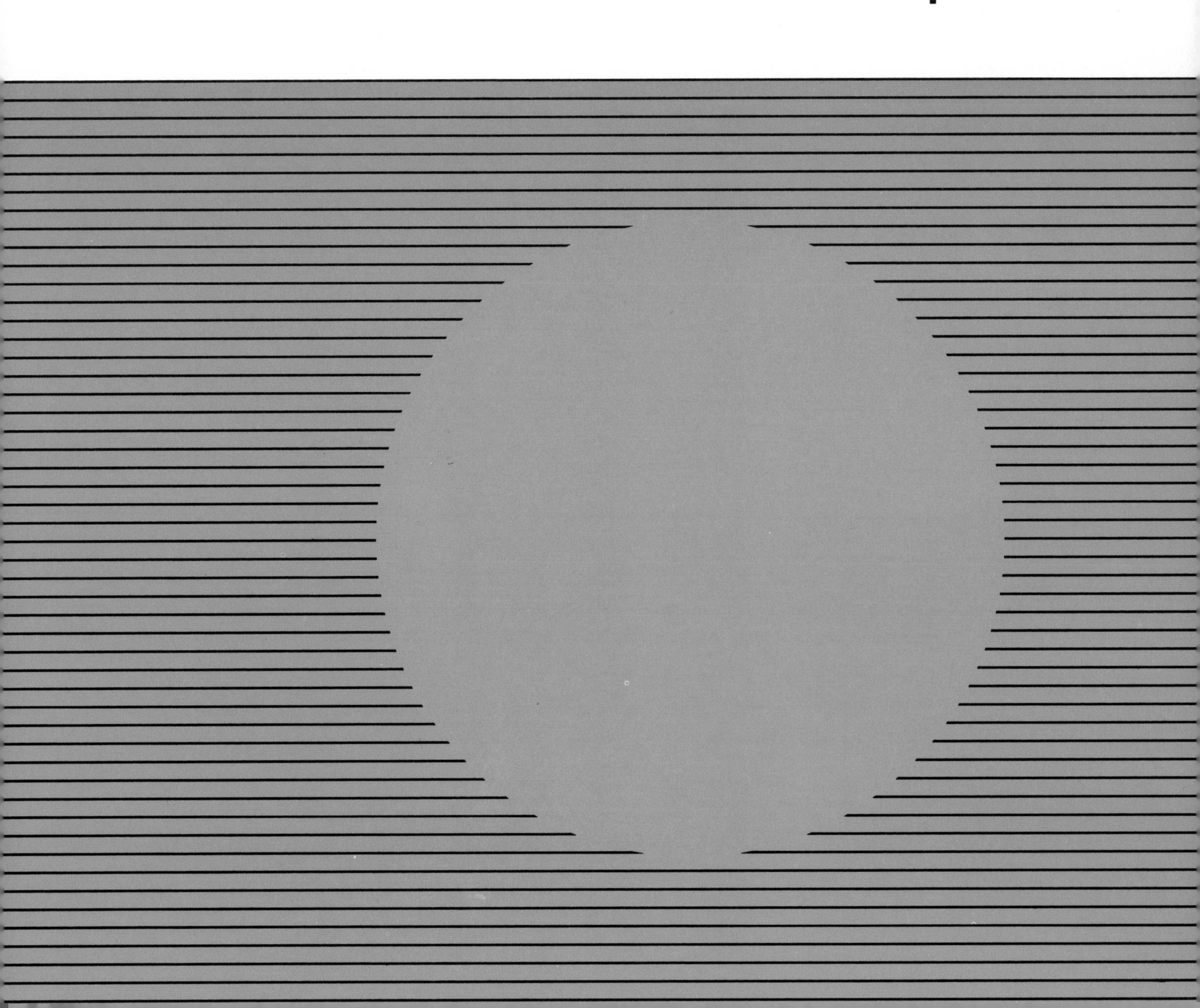

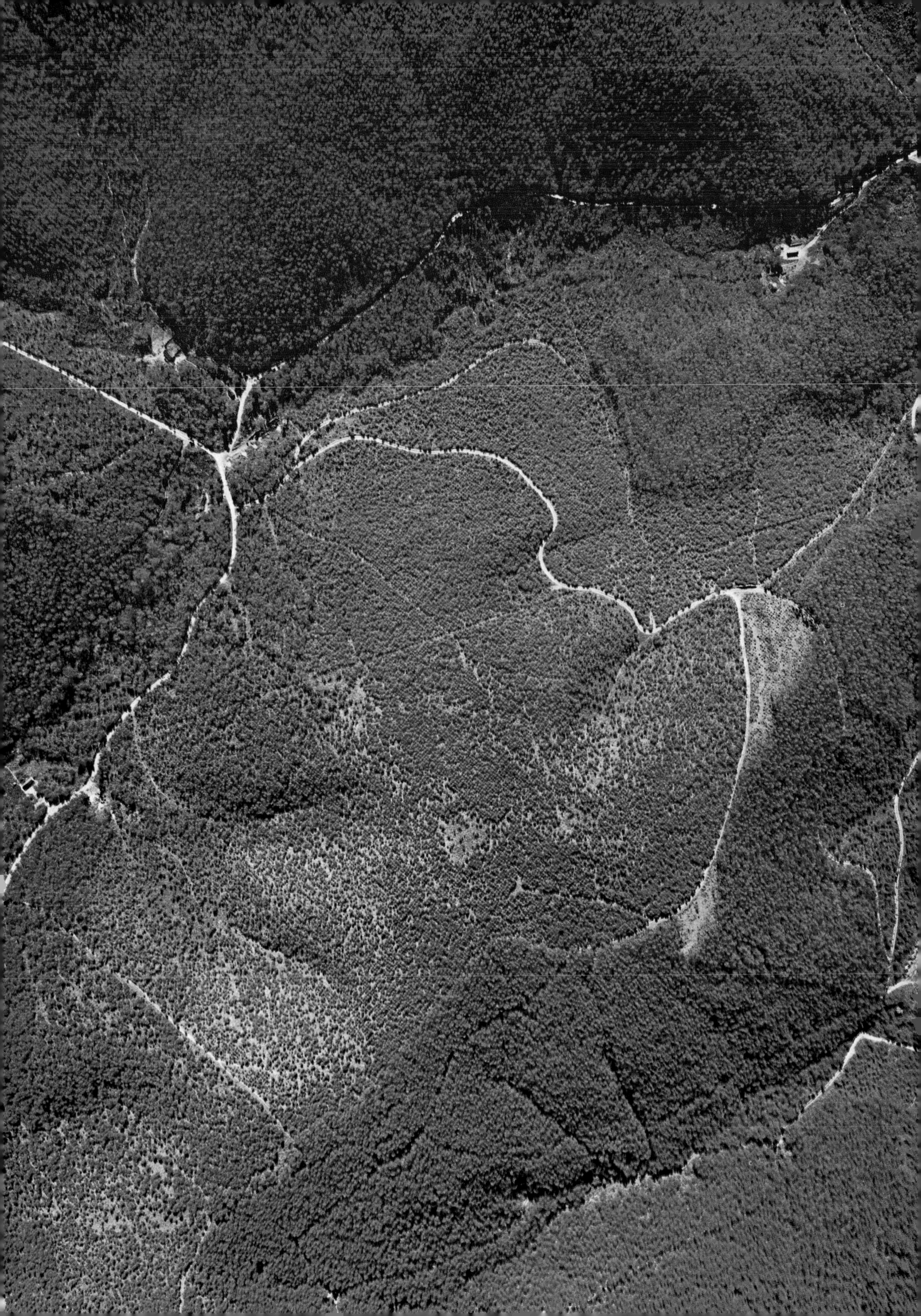

The Flow of Energy in the Biosphere

DAVID M. GATES

The solar energy that falls on the earth warms the surface and is ultimately radiated back into space. The tiny fraction of it that is absorbed by photosynthetic plants maintains all living matter

The radiant energy that bathes the earth builds order from disorder through the processes of life. Most events in the universe proceed toward increasing entropy, but life postpones the effect of this basic law by using the stream of sunlight to build highly complex assemblages of proteins, carbohydrates, lipids and other biological molecules. The aim of this article is to trace the radiant energy as it sustains the remarkable diversity of living organisms.

A living organism can be viewed as a chemical system designed to maintain and replicate itself by utilizing energy that originates with the sun. Life cannot be sustained merely by an adequate quantity of radiation; the light must also be of a suitable spectral quality. The flux of solar radiation received at the ground is highly variable in both quantity and quality because of the variable transmissivity of the atmosphere and the changing degree of cloudiness. The earth's atmosphere filters sunlight by absorbing most of the ultraviolet wavelengths and some of the infrared.

Light entering a chemical system can be utilized in several ways. It can be absorbed and then simply dissipated as heat through the increased motion of the molecules in the system. It can be reradiated at the resonant frequencies of the molecules or as fluorescence or phosphorescence. It can be utilized to accelerate a chemical reaction that either increases or decreases the free energy of the participating molecules.

Light consists of the bundles of energy called quanta. The energy content of a quantum is proportional to the frequency of the light: the shorter the wavelength, the higher the frequency and the greater the energy content. A mole of any substance (a weight in grams equal to the molecular weight of the substance) contains 6×10^{23} molecules; that is the universal constant known as Avogadro's number. In discussing the interaction of light and matter it is convenient to use one mole of a substance. The energy content of a molecular bond can then be multiplied by the number of molecules per mole (6×10^{23}) to get the bond energy of the substance per mole. One can also regard one mole as containing 6×10^{23} quanta and can multiply that by the energy per quantum in order to get the mole equivalent energy of radiation.

The mole equivalent energy of blue light at a wavelength of 450 nanometers (a nanometer is a billionth of a meter) is 64 kilocalories per mole; of infrared radiation at 900 nanometers, 32 kilocalories per mole, and of ultraviolet radiation at 225 nanometers, 128 kilocalories per mole. The strength of molecular bonds (or the energy required to break them) can be expressed in kilocalories per mole. A single bond between two carbon atoms can be broken with only 82.6 kilocalories per mole; a double bond between the two atoms requires 145.8 kilocalories per mole and a triple bond 199.6 kilocalories per mole.

It is evident from these numbers that ultraviolet radiation has the energy per mole necessary to break bonds. It is also clear that visible light has relatively little potential for breaking or forming bonds and that infrared radiation has even less. Light absorbed by a molecule kicks one of the electrons associated with the molecule into an excited energy state, thereby making the electron available for pairing with an electron from a neighboring atom or molecule in an electron-pair bond. By this photochemical process new molecules are formed.

The most fundamental photochemical reaction of life is photosynthesis in plants. Photosynthesis combines molecules of carbon dioxide and water to form carbohydrate and oxygen; the energy converted in the process is 112 kilocalories per mole. It is known that photosynthesis proceeds by means of blue light and red light. From the relation of energy and wavelength I have described it is clear that neither blue nor red light can directly provide enough energy for photosynthesis.

It turns out that photosynthesis is a complicated stepwise process. Light is absorbed by the chlorophyll molecule (and by other pigments in the plant) and is transferred to electrons in such a way as to create strong oxidants and reductants, that is, molecules that readily remove electrons from other molecules (oxidize them) or readily supply electrons to other molecules (reduce them). In photosynthesis the oxidants and reductants assist with the storage of energy in chemical bonds, notably those of carbohydrate and of adenosine triphosphate (ATP), the basic energy currency of all living cells.

Animals, by eating plants, are able to release the energy stored in them by

ABSORPTION AND REFLECTION characteristics of vegetation are partly indicated by the aerial photograph on the opposite page. The vegetation, which is in a forested area northwest of São Paulo in Brazil, is red because the photograph was made with a special emulsion that is sensitive in the near-infrared. Green plants absorb about 92 percent of the blue and red light that energizes the process of photosynthesis. They absorb some 60 percent of the near-infrared; the rest is reflected. Thus a photograph of vegetation in the near-infrared shows considerably more intense reflection than one in visible region of spectrum.

means of the various oxidative reactions of metabolic processes. ATP interacts with the carbohydrate glucose to prepare it, through glycolysis, for a long series of complex reactions in the metabolic sequence known as the citric acid cycle. The energy released is employed to do muscular work, to generate nerve impulses and to synthesize proteins and other molecules for the building of new cells. The entire chain of life proceeds in this way as energy cascades through the communities of plants and animals.

Living systems must be protected against an excess of bond-breaking radiation. The primordial atmosphere of the earth contained no free oxygen and was highly transparent to ultraviolet radiation. Once photosynthesis began (with microorganisms in the ocean) oxygen was released to the atmosphere. Since the metabolic processes of the primitive organisms were primarily anaerobic, the oxygen in the atmosphere built up. As the oxygen molecules (O_2) diffused upward they were decomposed by ultraviolet radiation into oxygen atoms (O), some of which formed ozone (O_3). Ozone strongly absorbs ultraviolet radiation, and as it built up in the stratosphere it acted as a filter. In this way the earth's surface was shielded against the energetic ultraviolet radiation of the sun but remained transparent to visible light.

The interaction of life and the atmosphere has many more components than the ozone shield against ultraviolet. For example, the atmosphere contains .032 percent, or 320 parts per million, of carbon dioxide, which is essential to the part of photosynthesis that assimilates carbon into carbohydrates. Carbon dioxide, which at visible wavelengths is a clear transparent gas, strongly absorbs the radiation in certain infrared bands of the spectrum. The earth's surface radiates heat into space entirely at infrared wavelengths. If there were no atmosphere, or if the atmosphere were fully transparent, the temperature of the ground at night would be considerably colder than it is.

What happens is that the absorption bands of atmospheric carbon dioxide capture some of the infrared radiation headed toward space from the earth. The captured energy is then reradiated in two directions: back toward the ground and into outer space. Therefore the ground is exposed not to the cosmic cold of outer space but to a warm flow of radiation emitted by atmospheric carbon dioxide.

Clouds and water vapor in the sky also absorb and emit infrared radiation. When the sky is clear, the water vapor and the carbon dioxide only partly

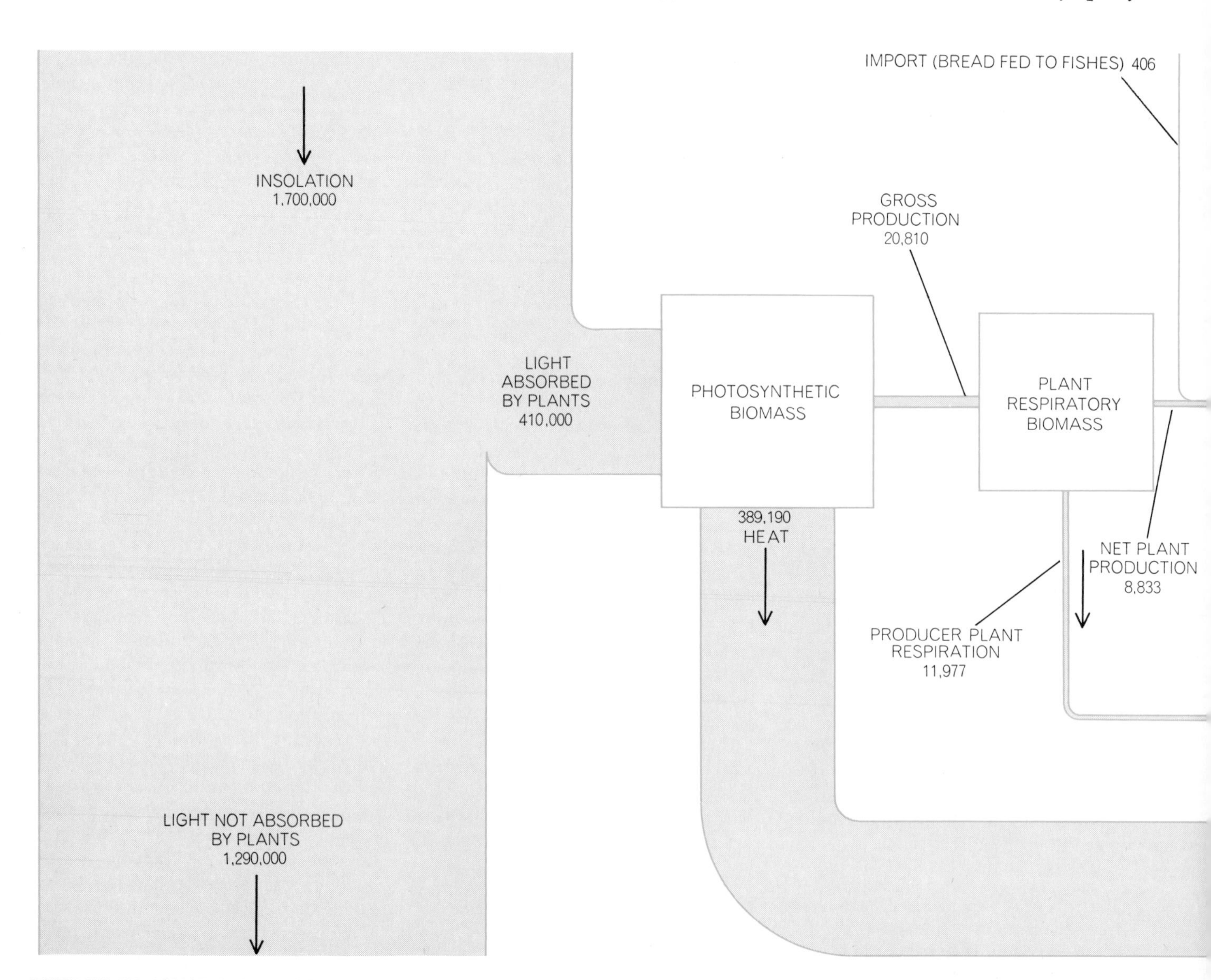

ENERGY IN A NATURAL SYSTEM flows as indicated in this diagram of the ecosystem at Silver Springs, Fla., which consists of a clear, spring-fed stream with vegetation covering the bottom and numerous species of animals living in or near the water. The nu-

shield the ground from the cold of space. When the sky is overcast, the cloud cover serves as an opaque thermal blanket. At such times the radiation from the earth to space originates at the top of the cloud deck rather than at the ground.

Thus green plants not only get the benefit of carbon dioxide but also are warmed by the radiant flux returned to the ground from the atmosphere. The atmosphere's window on space is transparent to visible light but is closed at the ultraviolet end by ozone absorption and at the infrared end by absorption in carbon dioxide and water vapor. This grand-scale synergy of green plants and the atmosphere is the result of millions of years in evolution of life and of the atmosphere, which are therefore closely interdependent. Life depends on both the clarity and the opaqueness of the atmospheric window, and the waste products of man that are discharged into the atmosphere dirty the window, on whose clarity all life depends.

The growth of green vegetation depends simultaneously on the amount of sunlight reaching the ground, the temperature near the surface and the amount of water available. If any of these conditions is inadequate, growth is reduced. Much sunlight and little water produce a desert. Much sunlight and low temperature produce tundra. Much water and little sunlight make for a stunted rain forest.

The annual productivity of green vegetation is limited by the seasonal distribution of sunlight, temperature and moisture. The solar radiation reaching the atmosphere is partly absorbed by ozone, carbon dioxide, water vapor, nitrogen, oxygen, dust and aerosols. By the time it reaches the ground it is weakened in intensity and modified in spectral quality. Solar radiation at the ground—direct sunlight plus skylight—varies from a maximum of between 200 and 220 kilocalories per square centimeter per year in desert areas to 70 kilocalories per square centimeter per year in polar regions. Tropical rain forests receive from 120 to 160 kilocalories; much of Europe, 80 to 120. The solar radiation at the Equator varies relatively little during the year except as it is affected by cloudiness. Polar regions experience the midnight sun of summer and perpetual darkness during the winter.

Now that we have traced the solar flux down through the atmosphere to the surface of the ground, let us see how it is partitioned and what it does. Clearly if it strikes bare rock or soil, it will be partly reflected and partly absorbed, and the rock or soil will gain energy. If the surface bears vegetation,

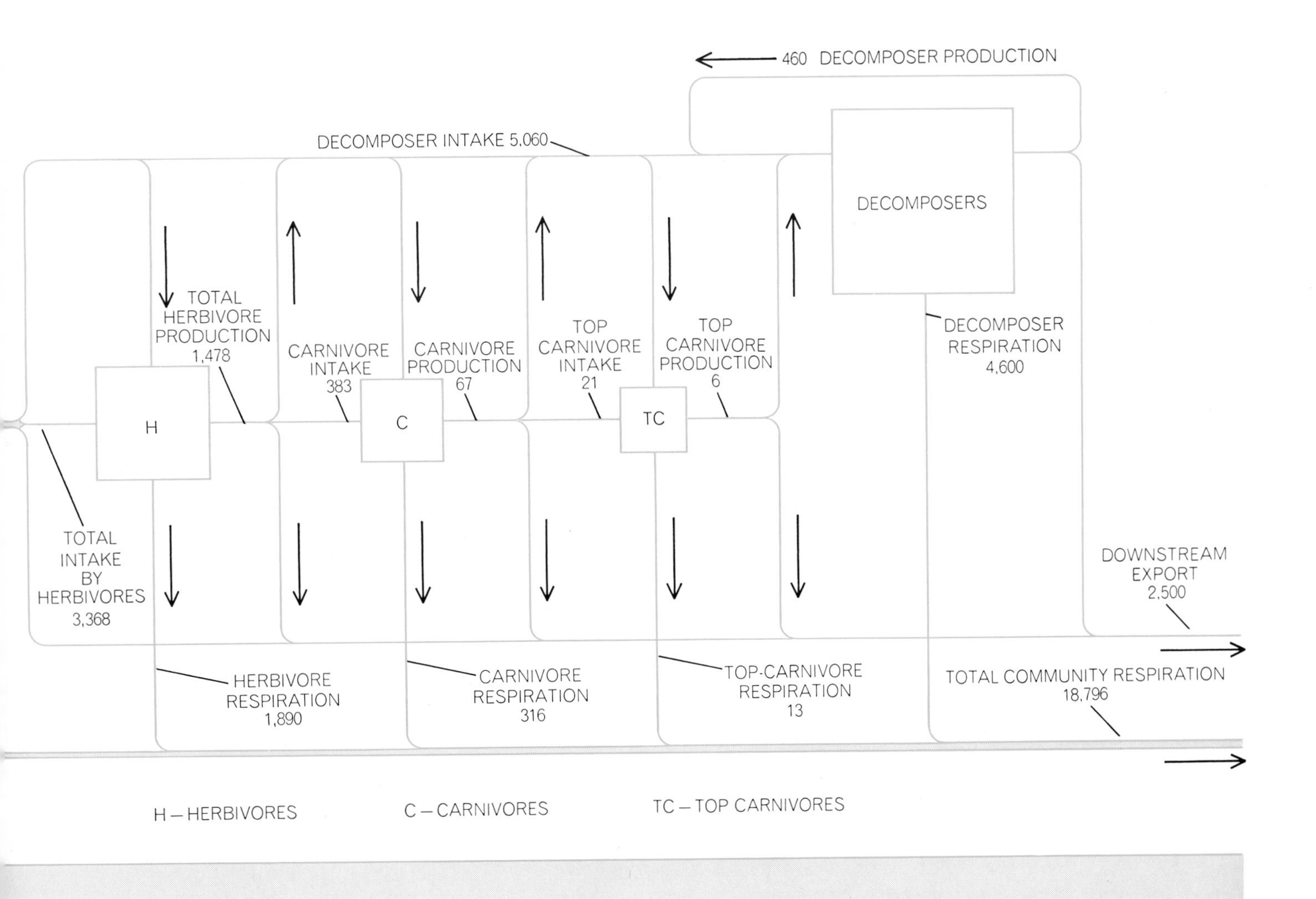

merals give inputs and outputs in kilocalories per square meter per year, with relative energies indicated by the widths of the bands and lines. The data were obtained by Howard T. Odum of the University of Florida. Top carnivores are at top of food chain.

some of the incident solar radiation is utilized in photosynthesis. Green fields reflect from 10 to 15 percent of visible light; dark green coniferous forests may reflect only 5 to 10 percent.

Of the total amount of solar energy entering the earth's atmosphere only about 53 percent is available at the ground after all scattering, absorbing and reflecting processes are taken into account. The ground exchanges energy by radiation, by the evaporation and condensation of water, by the exchange of sensible heat between the surface and the air and by conduction into or out of the soil. All the energy flowing to or from the ground must be accounted for in the energy budget relating to the surface.

During the day the surface has a net influx of radiation; during the night it loses a net quantity of radiation. During the day, when the ground is warmer than the air, heat is transferred from ground to air by convection. At night the air is usually warmer than the ground, so that the convectional transfer of heat is from air to ground.

Evaporation of water away from the surface requires both a moisture gradient away from the surface and energy sufficient to supply the latent heat of vaporization. The amount of energy required is about 580 calories per gram at a temperature of 30 degrees Celsius. Evaporation and evapotranspiration through the leaves of plants are almost always taking place in daylight, except when it is raining; sometimes they proceed at night as well.

If the ground is quite dry, the net radiation input during the day will go into convection and conduction. The environment will be turbulent and windy, as is typical of deserts. If the ground is moist or the vegetation is well watered, evapotranspiration will consume the major fraction of net radiation and the atmosphere will be more quiescent. These are basically physical processes related to the thermodynamics of the earth's surface and to the conditions of climate. Let us now leave them in order to trace out the pathway of light—the visible wavelengths of radiation—as it affects primary productivity (the growth of vegetation) and the food chain of life.

Of the total amount of sunlight reaching the ground only about 25 percent is of wavelengths that stimulate photosynthesis, and only a fraction of the 25 percent is actually used by green plants. Most plants in the open are using light at their maximum rate during most of the hours of daylight. A forest or a field receives, on a typical summer day in the U.S., from 500 to 700 calories of solar energy per square centimeter per day. Assuming that plants are receiving such an input and that they are using as much sunlight as they can, one can estimate the productivity of growing things.

Robert S. Loomis and William A. Williams of the University of California at Davis have made such estimates. Considering 500 calories per square centimeter per day a typical daily input of energy during the growing season, they found that potential net plant production (gross production minus respiration) is about 71 grams per square meter per day. Assuming that this net productivity represents as storage in carbon compounds about 3,740 calories per gram, one finds that 26.6 calories per square centimeter per day of solar radiation ends up in biomass. This represents 5.3 percent of the total incident solar radiation and about 12 percent of the energy received as visible light, which is 222 calories per square centimeter per day.

Plants reflect about 8 percent of photosynthetically active wavelengths. The

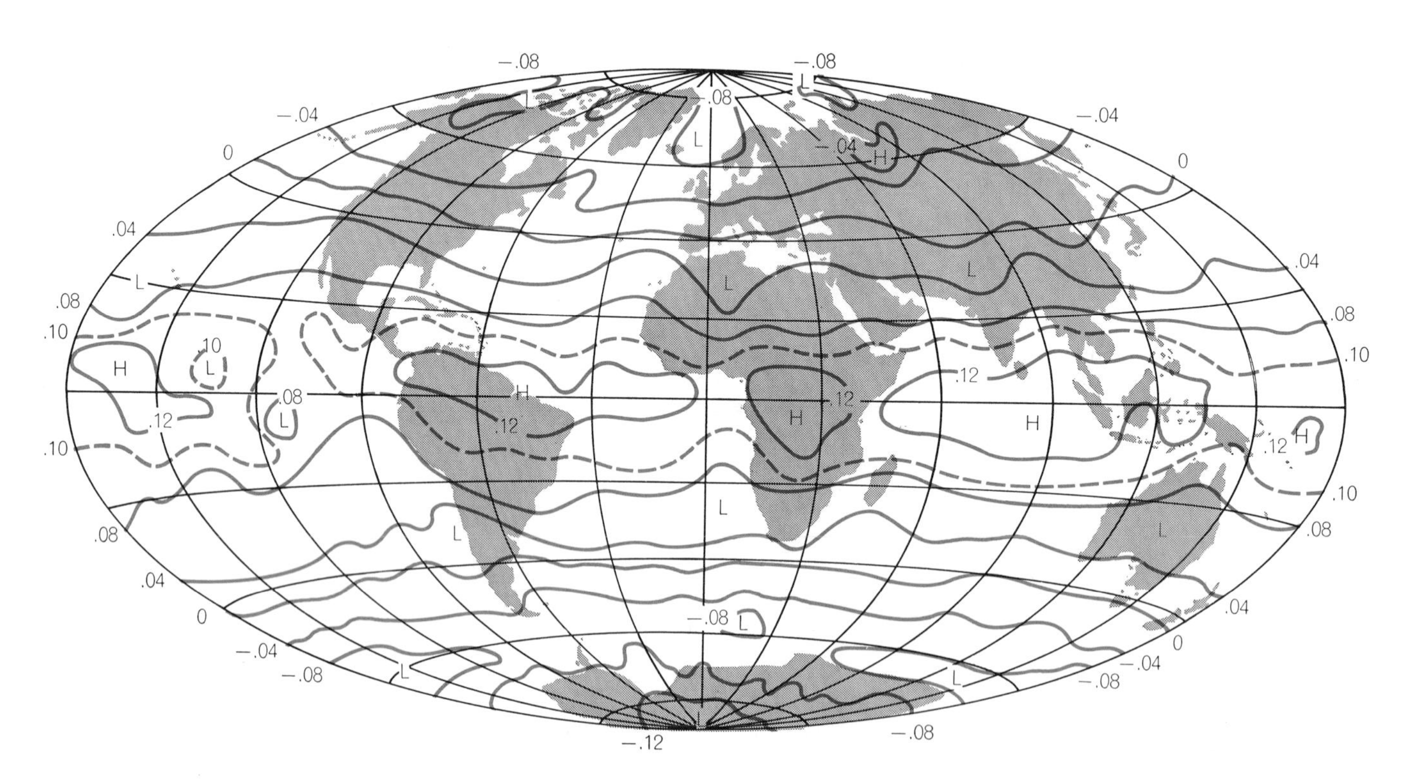

MEAN RADIATION of the earth is portrayed by isopleths (*color*) that give the net radiation in terms of calories per square centimeter per minute. Areas marked *H* and *L* are respectively high or low compared with their surroundings. The data were obtained by satellites that measured the earth's albedo, which indicates how much of the solar radiation reaching the earth is reflected and how much is absorbed, and also measured the long-wave radiation from the earth. The isopleths of the map give the resulting net radiation and thereby provide information about the exchange of energy between the earth and space. The work was done in the department of meteorology of the University of Wisconsin by Thomas H. Vonder Haar, now at Colorado State University, and Verner E. Suomi.

percentage is considerably higher for the portion of such wavelengths in the near-infrared region of the spectrum; there individual leaves have reflectivities of 40 percent or more. An individual leaf will absorb some 60 percent of the total incident sunlight. A dense stand of vegetation, however, will absorb considerably more. For example, a dense corn crop may reflect 17 percent of the total incident solar radiation, transmit to the soil about 13 percent and absorb in the leaves about 70 percent.

About 10 quanta of light are required to reduce one molecule of carbon dioxide to carbohydrate. Respiration consumes from 20 to 40 percent of gross photosynthesis; the value used by Loomis and Williams for their estimates of productivity was 33 percent. Gross productivity was 107 grams per square meter per day and respiration burned up 36 grams per meter per day. Of the 500 calories per square centimeter per day incident on the crop 375 calories were absorbed by the crop and the soil, converted to heat and then transferred by radiation to the atmosphere, by evapotranspiration, by convection to the air and by conduction into the soil. The evapotranspiration component may account for as much as 200 calories per square centimeter per day.

A crop of Sudan grass at Davis produced 51 grams per square meter per day during a 35-day period when the incident solar radiation averaged 690 calories per square centimeter per day. It is estimated that the maximum potential production by this crop was 104 grams per square meter per day, so that the actual production was 49 percent of the potential yield. At 51 grams per square meter per day and with a caloric content of 4,000 calories per gram, the crop put into storage 3 percent of the total incident solar radiation and 6.7 percent of the visible radiation. Barley has been observed to convert as much as 14 percent of the incident visible light to carbon compounds. In general the productivity of crops is much lower than these maximum values.

Jen-hu Chang of the University of Hawaii has made careful estimates of potential photosynthesis and crop productivity for various regions of the world. He based his estimates on the intensity and duration of sunshine and the mean monthly temperatures over the period under consideration, and he assumed a well-watered crop. He made estimates of potential net photosynthesis, expressed in terms of crop productivity in grams per square meter per

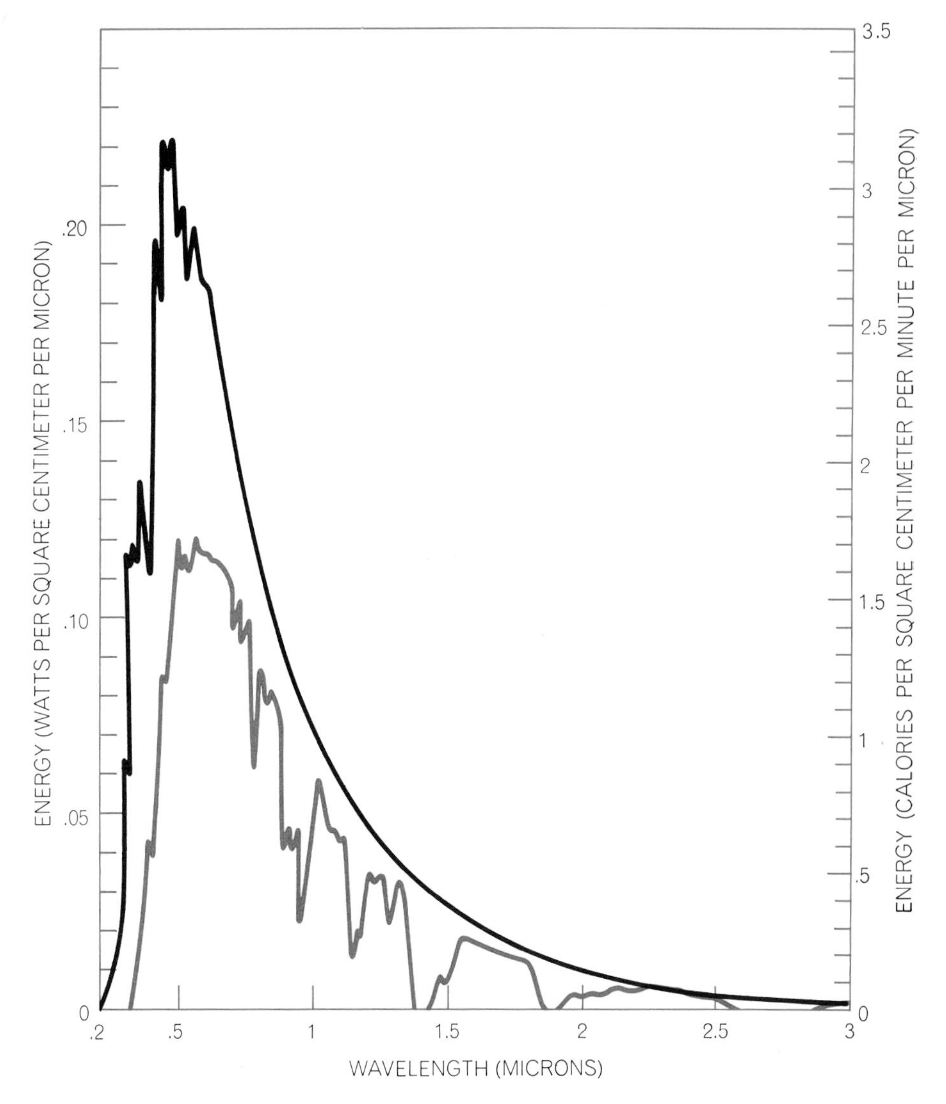

SPECTRAL DISTRIBUTION of solar radiation reaching the earth is given for the top of the atmosphere (*black*) and the ground (*color*). Curve for ground takes into account the absorbing effects of water vapor, carbon dioxide, oxygen, nitrogen, ozone and particles of dust. Data are based on a solar constant of 1.95 calories per square centimeter per minute.

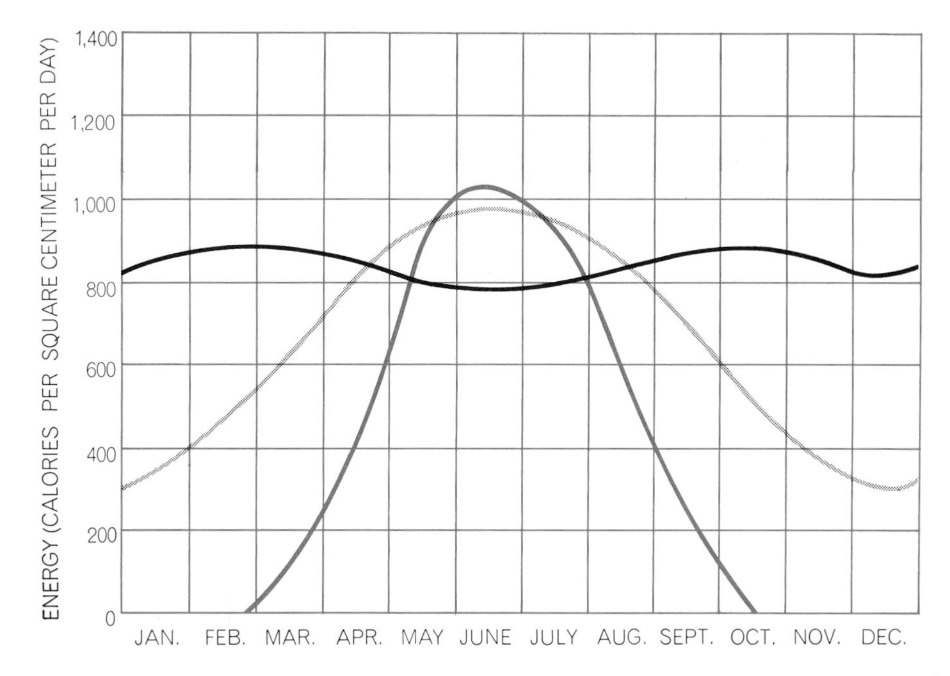

SEASONAL VARIATION of solar radiation on a horizontal surface outside earth's atmosphere is given for Equator (*black*) and latitudes 40 degrees (*gray*) and 80 degrees (*color*).

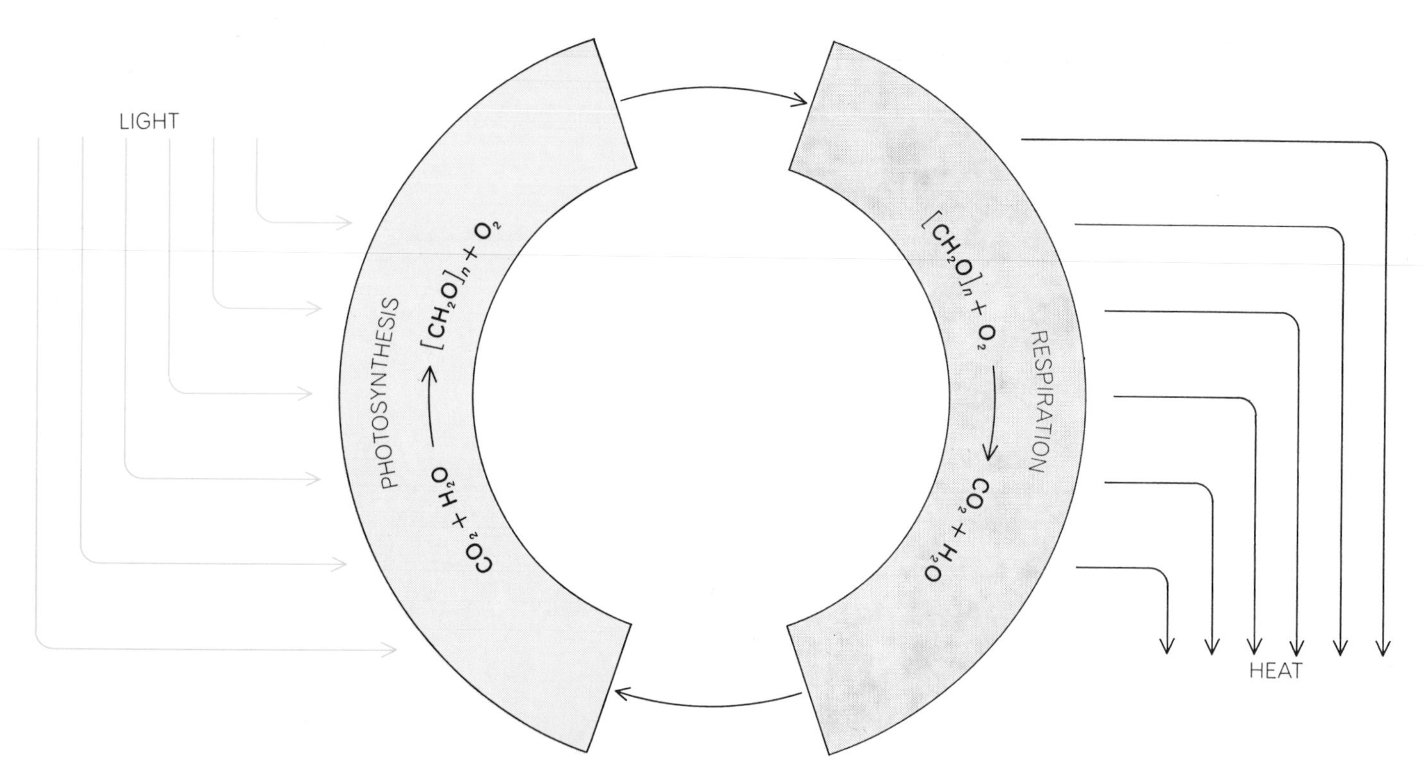

PHOTOSYNTHESIS AND RESPIRATION, which are the basic metabolic processes of most living plants, obtain their energy from sunlight. In photosynthesis energy from light is used to remove carbon dioxide and water from the environment; for each molecule of CO_2 and H_2O removed, part of a molecule of carbohydrate (CH_2O) is produced and one molecule of oxygen (O_2) is returned to the environment. In respiration by a plant or an animal combustion of carbohydrates and oxygen yields energy, CO_2 and H_2O.

day, for a four-month summer period and an eight-month period centered on the summer; he also calculated an annual mean [*see illustration below*]. His results are of much interest and should be compared with the highest levels of plant production discussed above.

During the four-month summer period the lowest potential photosynthesis is in the Tropics, and in particular at 10 degrees north latitude, which is the heat equator of the earth. There the potential photosynthesis is 25 percent lower than it is in temperate regions. Highlands in the Tropics have a higher yield than hot, humid lowlands. Southern Alaska, the upper Mackenzie River region in northwest Canada, southern Scandinavia and Iceland have the highest potential photosynthesis, exceeding 37.5 grams per square meter per day. The reason is that these regions receive many hours of sunlight per day during the summer. Farther north the effect of low temperature drops productivity even though the summer day is still longer.

Across the central U.S. the potential net photosynthesis is about 30 grams per square meter per day for both the four-month period and the eight-month period. At the Canadian border the

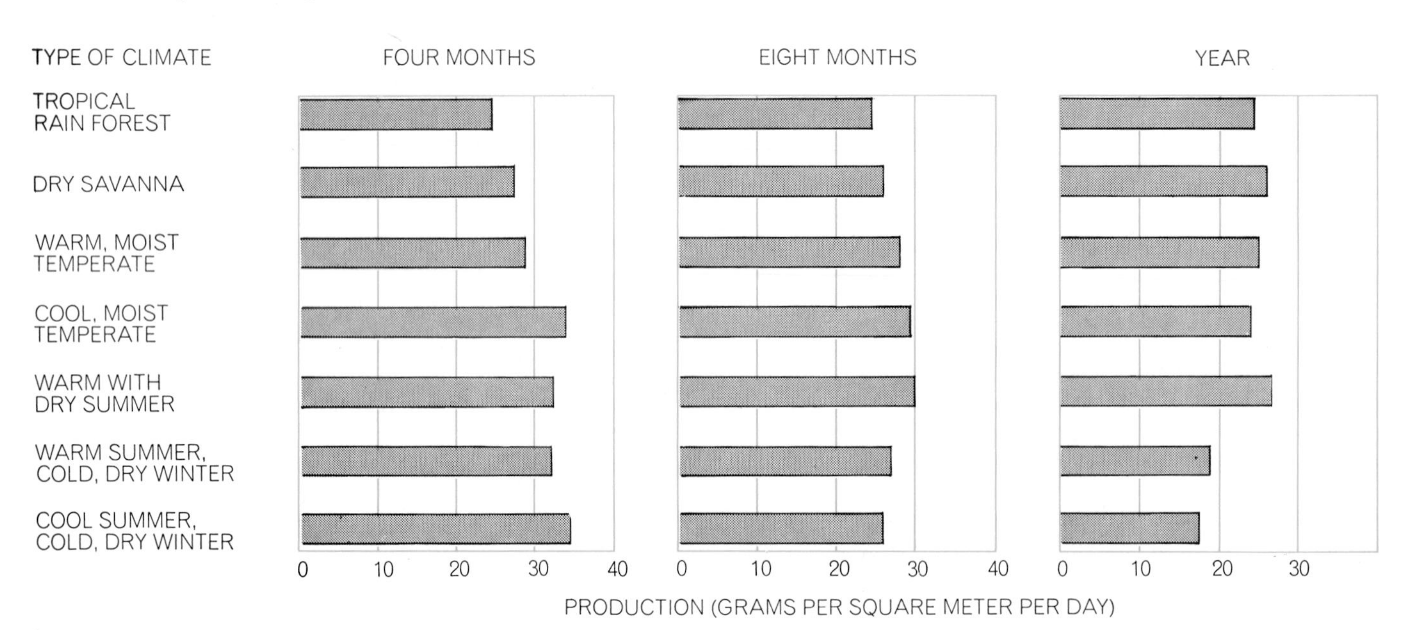

POTENTIAL NET PHOTOSYNTHESIS is expressed in terms of plant productivity in grams per square meter per day for various climates. Calculations are made for a four-month period, representing the normal growing season for most plants, for an eight-month period centered on the summer and for a year. Net production is found by subtracting respiration from gross production.

four-month value is 36 grams per square meter per day, the eight-month value 27.5 grams per square meter per day and the annual value 17.5 grams per square meter per day. The annual value through the central U.S. is from 20 to 22.5 grams per square meter per day. The same levels are found throughout much of Europe, except in Spain, where the level is above 25 grams per square meter per day. Northern Europe has enormous four-month potential photosynthesis, with values of up to 38 grams per square meter per day. Clearly these high-latitude regions are best suited to crops with a short growing season. It is significant in this context that the part of western Europe between 50 and 60 degrees north latitude leads the world in wheat production.

Chang has made a number of highly interesting assessments of actual yields compared with potential photosynthesis in various countries. He finds that all the developed countries have a four-month potential photosynthesis in excess of 27.5 grams per square meter per day, whereas all the underdeveloped countries have much lower values. What the difference means is that countries such as the Philippines can expect to increase their yields by 30 percent through improved agricultural methods, but no matter what they do they cannot improve 500 percent or more in order to reach the yields of such countries as Spain. In other words, the underdeveloped countries are climatically limited. Chang has compared rice yields with the four-month potential photosynthesis, cotton yields with the eight-month potential (since cotton is planted in early spring and harvested in late fall) and sugarcane yields with the annual potential values (because sugarcane has a long growing season). The situation is the same throughout: the underdeveloped countries are the climatically deprived countries.

It is useful to compare the maximum rates of photosynthesis by agricultural crops, which are in principle plants selected and cultivated for high productivity, with the levels of productivity achieved by natural plant communities. On an annual average the net productivity in grams per square meter per day was nine for spartina grass in a salt marsh in Georgia; six for a pine forest in England during the years of most rapid growth; three for a deciduous forest in England; 1.22 for tall-grass prairies in Oklahoma and Nebraska; .19 for a short-grass prairie in Wyoming, and .11 for a desert in Nevada with five

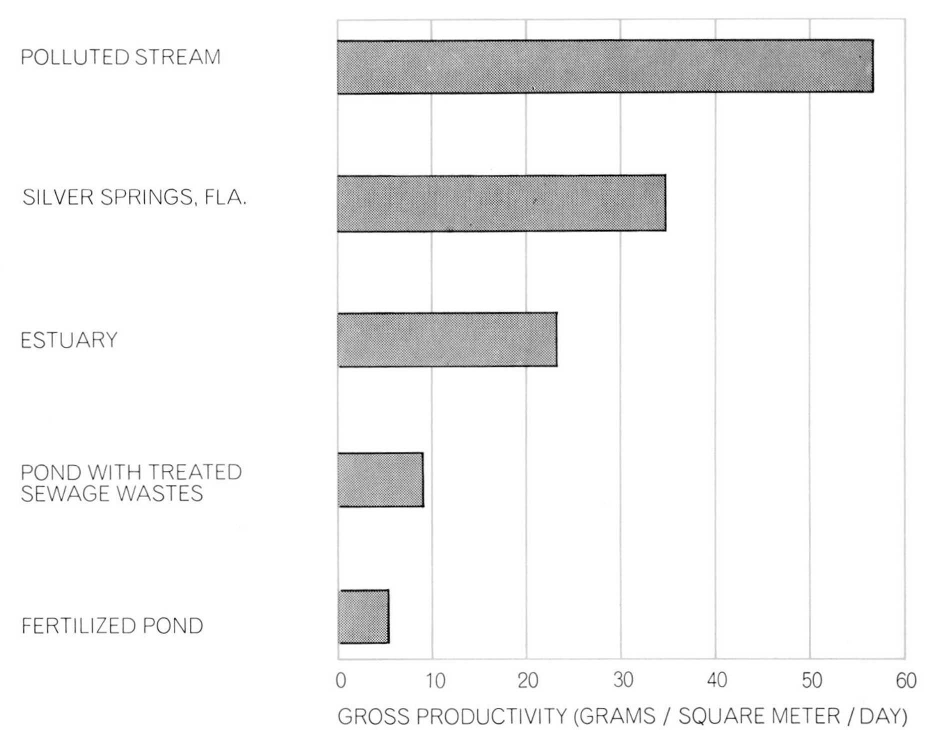

GROSS PRODUCTIVITY in several environments during short periods of time favorable for growing is expressed in terms of grams of dry matter produced per square meter per day. The polluted stream was in Indiana, the estuary was one of several in Texas, the pond with treated sewage wastes was in Denmark and the fertilized pond was in North Carolina.

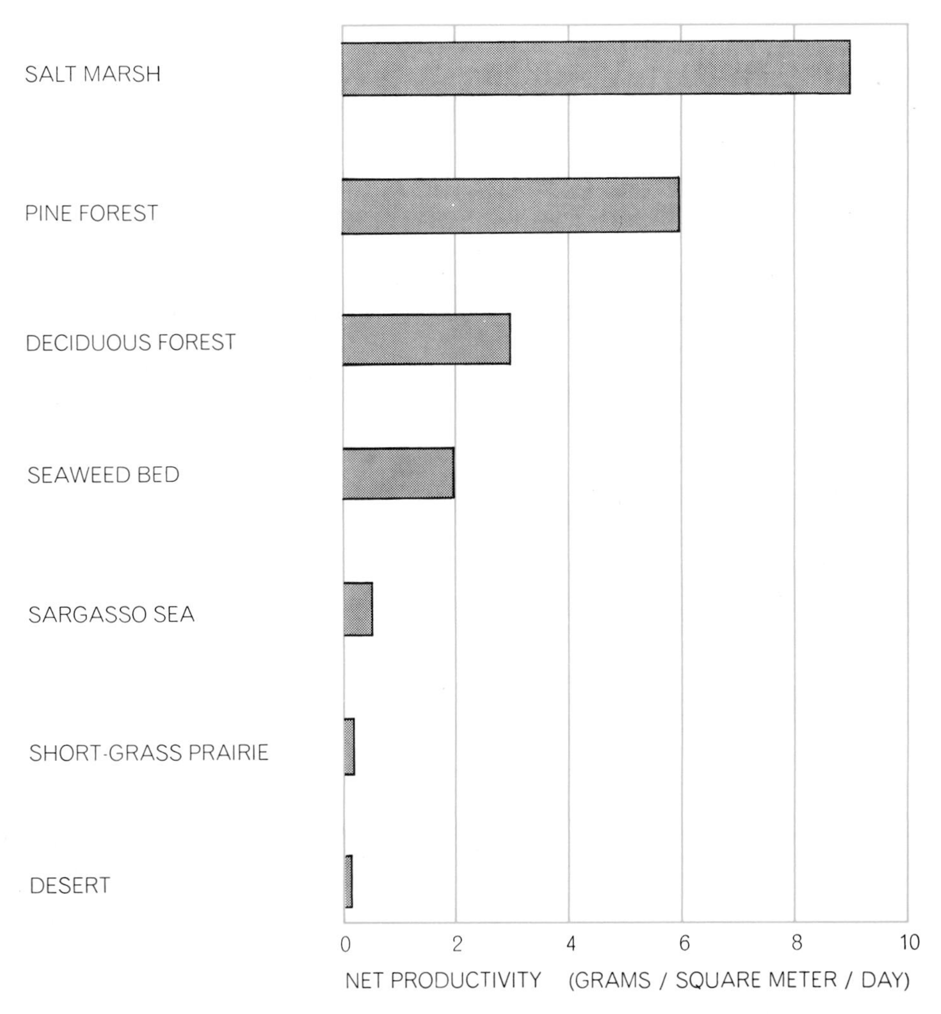

NET PRODUCTIVITY of seven environments is depicted. The salt marsh was in Georgia, forests were in England, prairie and desert in the U.S. and seaweed beds in Nova Scotia.

inches of rain per year. It is evident that the net productivity of most natural stands of vegetation is considerably lower than the levels achieved for crops managed by man.

Since each gram of dry matter produced has a caloric value of about 4,000 calories, the productivities can be converted into percentages of solar radiation utilized. For example, the salt marsh in Georgia converted nine times 4,000 calories per square meter per day into dry matter from the incident solar radiation of about 4.5 million calories per square meter per day for a conversion factor of .8 percent. The conversion by a forest is about .5 percent, by a tall-grass prairie about .1 percent and by a desert .05 percent or less. The average cornfield utilizes about 1 percent of the incident solar radiation and seldom exceeds 2 percent.

Many factors can limit primary productivity. If only one of them is greatly different from the optimum, productivity decreases. Among the important factors are sunlight, carbon dioxide, temperature, water, nitrogen, phosphorus and trace amounts of several minerals. A tropical forest may have an enormous standing mass of vegetation, but the rate of growth is not as high as in other regions because of high temperatures and limitations of soil nutrients. The primary productivity of deserts and grasslands is severely limited for lack of water. Alpine and arctic tundras, which are very wet because of low evaporation and the presence of permafrost not far below the surface, are limited in productivity by low temperatures.

All vegetation is limited in growth by the concentration of carbon dioxide in the atmosphere or in bodies of water. Plants will increase photosynthesis with increasing concentration of carbon dioxide to at least three times the normal concentration of 12.5 nanomoles per cubic centimeter (.03 percent by volume).

All life on the earth depends in one way or another on primary productivity, that is, on the growth of vegetation. Herbivores feed on the carbohydrates and proteins generated by photosynthesis, and carnivores get their energy by feeding on the herbivores. Decomposing organisms feed on both plants and animals, so that the material ingredients of life are returned to the soil and the cycle continues.

Evolutionary events over the past three billion years have created on the earth some two million species of insects, perhaps a million species of plants, 20,000 species of fishes, 8,700 species of birds and almost numberless kinds of microorganisms. Together they form a continuum of life over the surface of the earth. The organisms of the world are all interdependent, forming a vast web of protoplasm through which matter cycles and energy flows.

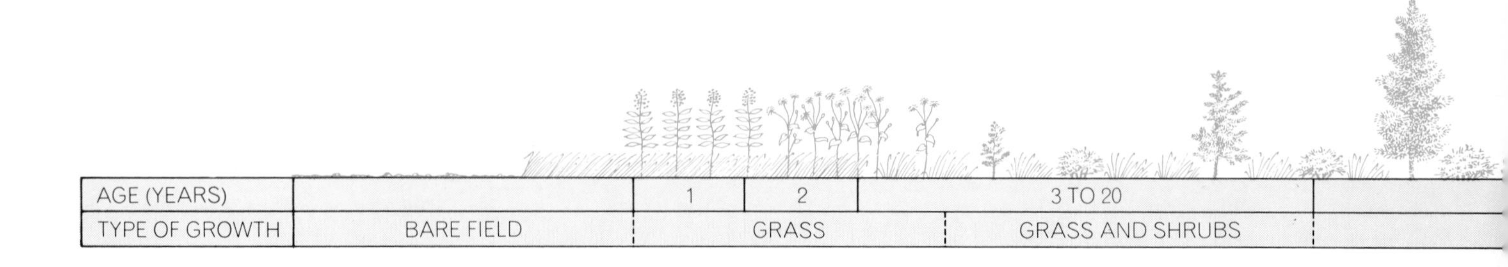

GROWTH SUCCESSION in the piedmont region of the southeastern U.S. progresses from grass to forest over a period of about 150 years. This is the succession that follows the abandonment of land once used for crops. Energy flows from a less mature ecosystem to a

The diversity of species within habitats varies enormously, from the incredible numbers of plant and animal species in a tropical rain forest to the severely limited varieties of life in a desert or a tundra. The diversity of species diminishes along gradients of water from moist to dry, of temperature from warm to cold and of light from bright to dark. By far the most intense speciation is found in a tropical rain forest.

Such a forest possesses an enormous biomass, consisting mostly of large trees laced with vines and lianas of various kinds struggling skyward from the dark interior to compete for light in the canopy. Tremendous numbers of insects and birds coexist in the canopy, and a constant rain of detritus falls to the forest floor, where decay returns the nutrients to the soil. Every conceivable niche of this biome is filled by a plant or an animal.

Biologists have often debated the reasons for the great diversity and stability of life in a rain forest. There is no single reason; there are several. The temperature is favorable to life and moisture is abundant. Physiological processes in animals proceed faster in warm climates than in cold ones. As a result life cycles are shorter and are repeated more often, so that genetic mutations are more likely to arise.

Perhaps the most significant feature is the relative constancy of the climate and the relative absence of fluctuation of environmental factors. For most animals there is a dependable supply throughout the year of the fruit, flower or seed that a particular species needs. The species can specialize in one or two kinds of food because it does not have to confront a lack of food at any time.

Mature and stabilized ecosystems tend to have less productivity than immature or transitional ones within the same general environment. Energy flows from the less mature one to the more mature one. The total amount of energy required to maintain a diverse and complex community of organisms is considerable, but per unit of biomass the amount of energy is smaller than it is in less complex communities.

A case in point is a grass meadow next to a forest. Left to itself the meadow will eventually become forested through a succession of plant communities beginning with grass and extending through herbs and shrubs to trees [*see illustration below*]. Compared with the forest, the meadow has high instability, low diversity of species and high productivity per unit of biomass. Excess energy is available. It is tapped by the plant succession, each member of which requires more energy than the preceding one, and by the animals of the forest, which feed on the plants and animals of the changing meadow. The natural

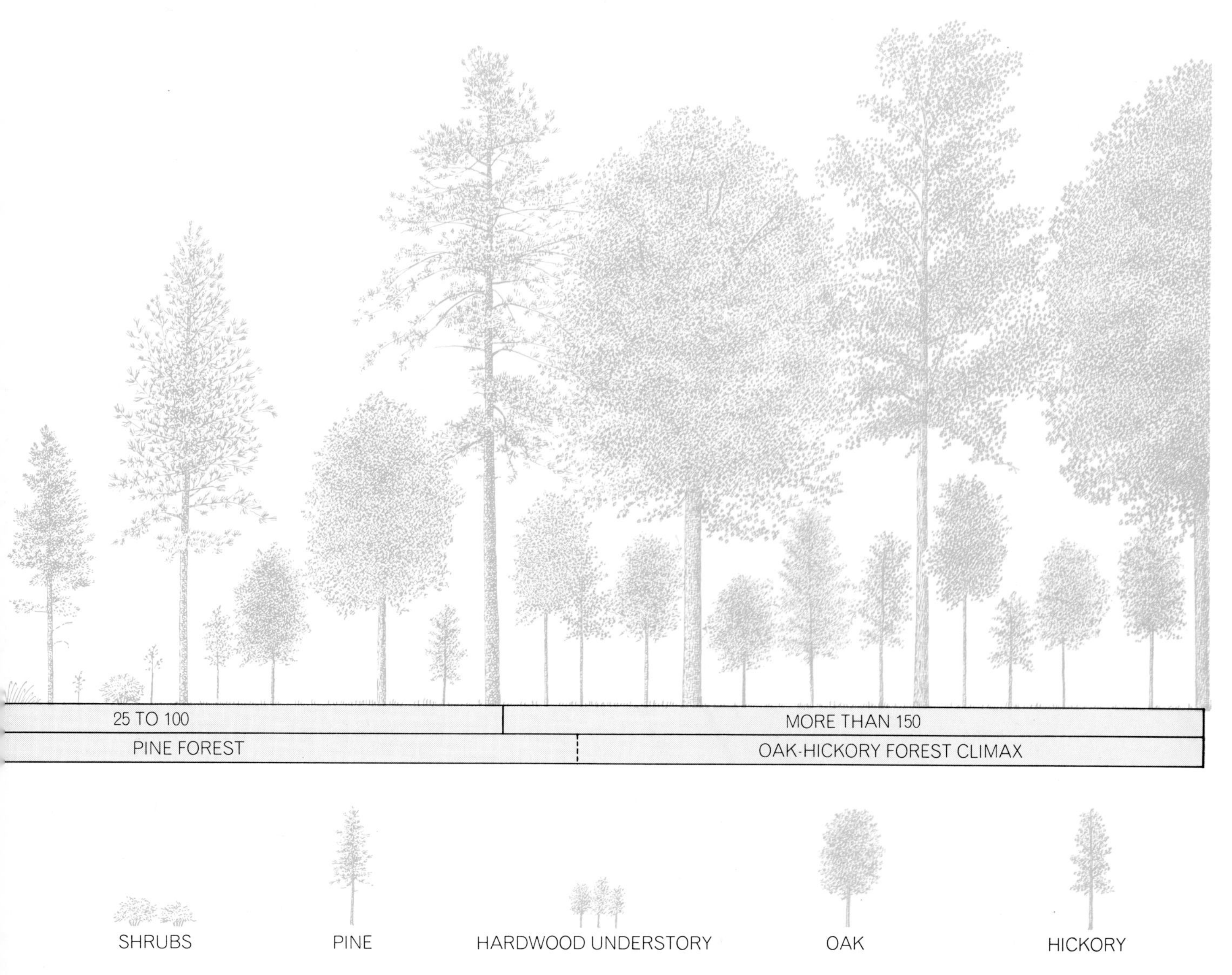

more mature one, which is to say that a mature, stabilized system is likely to have less productivity than an immature, transitional one.

The succession typical of abandoned farmland in the Southeast was ascertained by Eugene P. Odum of the University of Georgia.

trend in succession within communities is toward a decreasing flow of energy per unit of biomass and toward increasing organization.

In 1957 Howard T. Odum of the University of Florida published a thorough analysis of the flow of energy through a river in Florida: the famous tourist attraction of Silver Springs. It is of interest to follow the energy flow through this system [*see illustration on pages 44 and 45*]. During one year each square meter of the surface received 1.7 million kilocalories of solar energy. The green plants of the stream fixed 20,810 kilocalories per year in gross productivity, which represented an efficiency of 1.2 percent of the incident sunlight and 5.1 percent of the sunlight actually absorbed by the green plants. Respiration by the plants accounted for 11,977 kilocalories per square meter per year, so that the net productivity was 8,833 kilocalories per square meter per year. The herbivores converted 1,478 kilocalories per square meter per year into tissue while respiring 1,890 kilocalories. The carnivores had a net productivity of 73 kilocalories per square meter per year and respired 329. The efficiency of conversion to net productivity from the primary level to the secondary level, represented by the herbivores, was 18 percent; to the tertiary level, represented by the carnivores, it was 5 percent. The energy stored in the bodies of the carnivores was only one part in 23,300 parts of incident sunlight—a very small fraction indeed. Hence when man derives energy from a wild animal, he converts only a fraction of 1 percent of solar energy into body tissue. By domesticating plants and animals, however, he has considerably shortened the food chain and increased the efficiency of the total system.

Considering the relation in which man eats beef and beef animals eat corn (a food chain far commoner in the U.S. than in other countries), one can estimate the efficiency of a domesticated system. For simplicity the numbers that follow are only approximate. The cornfield can be considered to convert about 1 percent of solar energy. The beef animal will convert to body tissue about 10 percent of the energy stored in corn, and man will utilize about 10 percent of the energy stored in the tissue of the animal. Hence man derives at best about .01 percent of the incident solar energy through the food chain.

Man's basal metabolism is between 65 and 85 watts, depending on body size, or about .062 calorie per square centimeter per minute. An active adult walking slowly has a metabolic rate of 200 watts; if he walks rapidly, the rate rises to as much as 400 watts. On a daily basis man's minimum energy requirement is about 1,320 kilocalories if he is wholly sedentary. With moderate activity the need approaches 2,400 kilocalories per day. In cold climates the requirement rises to 3,900 kilocalories per day.

Assuming that the normal adult in our society needs 3,000 kilocalories per day, the requirement is equivalent to 30,000 kilocalories per day of beef, which in turn requires 300,000 kilocalories per day of energy in corn, which means 30 million kilocalories per day of sunshine. If the corn is produced in a region with an incident solar radiation of 500 calories per square centimeter per day, one can compute that it takes a cornfield with an area of 60 million square centimeters, or approximately 1.5 acres, to feed one person for one day by means of this food chain. The amount of land needed might be reduced somewhat by improved productivity, but at best it would not be less than one acre per day per person. How does this compare with the amount of land used to feed the world's population today?

Some 3.5 billion (3.5×10^9) acres are under cultivation, and some five billion more acres are used for grazing. That amounts to one acre of cultivated land and 1.5 acres of grazing land for each of the 3.5 billion people now living. The total of potential arable land is estimated at eight billion acres and of grazing land at an additional eight billion acres. Clearly in order to deliver 3,000 kilocalories per day to each person it would be possible to support only about 2.5 times the present population. Either we must increase the productivity per acre or substantially reduce the input per person. Many peoples of the world subsist on about 2,000 kilocalories per day (the global average is 2,350), but human beings at that level cannot be very energetic and cannot function effectively in a complex industrialized society.

This analysis is undoubtedly too simplistic. Even allowing for considerable improvement in productivity, higher levels of protein production, increased land use and extensive use of the oceans, however, it is unlikely that the earth could support more than 10 to 12 billion people reasonably well. If the global population never exceeds eight billion, the chances of feeding them well are much higher and the risks are greatly reduced.

5

The Flow of Energy in a Hunting Society

The Flow of Energy in a Hunting Society

WILLIAM B. KEMP

Early man obtained food and fuel from the wild plants and animals of his environment. How the energy from such sources is channeled is investigated in a community of modern Eskimos on Baffin Island

The investment of energy in hunting and gathering has provided man's livelihood for more than 99 percent of human history. Over the past 10,000 years the investment of energy in agriculture, with its higher yield per unit of input, has transformed most hunting peoples into farmers. Among the most viable of the remaining hunters are the Eskimos of Alaska, Canada and Greenland. What are the characteristic patterns of energy flow in a hunting group? How is the available energy channeled among the various activities of the group in order for the group to survive? In 1967 and 1968 I undertook to study such energy flows in an isolated Eskimo village in the eastern Canadian Arctic.

I observed two village households in particular. When I lived in the village, one of the two households was characterized by its "modern" ways; the other was more "traditional." I was able to measure the energy inputs and energy yields of both households in considerable detail. (The quantitative data presented here are based on observations made during a 54-week period from February 14, 1967, to March 1, 1968.) The different patterns of energy use exhibited by the two households help to illuminate the process of adaptation to nonhunting systems of livelihood and social behavior that faces all contemporary hunting societies.

For the Eskimos the most significant factor in the realignment of economic and social activity has been the introduction of a cash economy. The maintenance of a hunting way of life within the framework of such an economy calls for a new set of adaptive strategies. Money, or its immediate equivalent, is now an important component in the relation between the Eskimo hunter and the natural environment.

The village where I worked is one of the few remaining all-Eskimo settlements along the southern coast of Baffin Island on the northern side of Hudson Strait [*see illustrations on next page*]. In this village hunting still dominates the general pattern of daily activity. The economic adaptation is supported by the household routine of the women and is reflected in the play of the children. Villages of this type were once the characteristic feature of the settlement pattern of southern Baffin Island. Within recent years, however, many Eskimos have abandoned the solitary life in favor of larger and more acculturated settlements.

The community I studied is in an area of indented coastline that runs in a northwesterly direction extending from about 63 to 65 degrees north latitude. The land rises sharply from the shore to an interior plateau that is deeply incised by valleys, many containing streams and lakes that serve as the only routes for overland travel. At these latitudes summer activities can proceed during some 22 hours of daylight; the longest winter night lasts 18 hours. Perhaps the most noticeable feature of the Hudson Strait environment is tides of as much as 45 feet. Such tides create a large littoral environment; bays become empty valleys, islands appear and disappear and strong ocean currents prevail. In winter the tides build rough barriers of broken sea ice, and at low tide the steeper shorelines are edged with sheer ice walls. In summer coastal navigation and the selection of safe harbors are difficult, and in winter crossing from the sea ice to the land tries the temper of men and the strength of dogs and machines.

The varying length of the day, the seasonal changes of temperature, the tides and to some extent the timing of the annual freeze-up and breakup are predictable events, easily built into the round of economic activity. Superimposed on these events is the variability and irregularity of temperature, moisture and wind, which affect the pattern of energy flow for the community on a day-to-day basis. In winter the temperatures reach −50 degrees Fahrenheit, with a mean around −30. In summer the temperature may climb above 80 degrees, although temperatures in the low 50's are more typical. Throughout the year there may be large temperature changes from one day to the next. Midwinter temperatures have gone from −30 degrees to above freezing in a single night, bringing a thaw and sometimes even rain.

The heaviest precipitation is in the spring and fall, and strong winds can arise any day throughout the year. Winds of more than 40 miles per hour are common; on four occasions I measured steady winds in excess of 70 m.p.h. Speaking generally, the weather is most stable in March and April and least stable from late September into November.

Within this setting the Eskimos harvest at least 20 species of game. All the marine and terrestrial food chains are exploited in the quest for food, and all habitats—from the expanses of sea ice and open water to the microhabitats of

BLEAK TERRAIN of the Canadian Arctic is seen in the aerial photograph on the opposite page. The steep shore and treeless hinterland are part of an islet in Hudson Strait off the coast of southern Baffin Island. By hunting sea mammals the Eskimos of the region can obtain enough food and valuable by-products to keep them well above the level of survival.

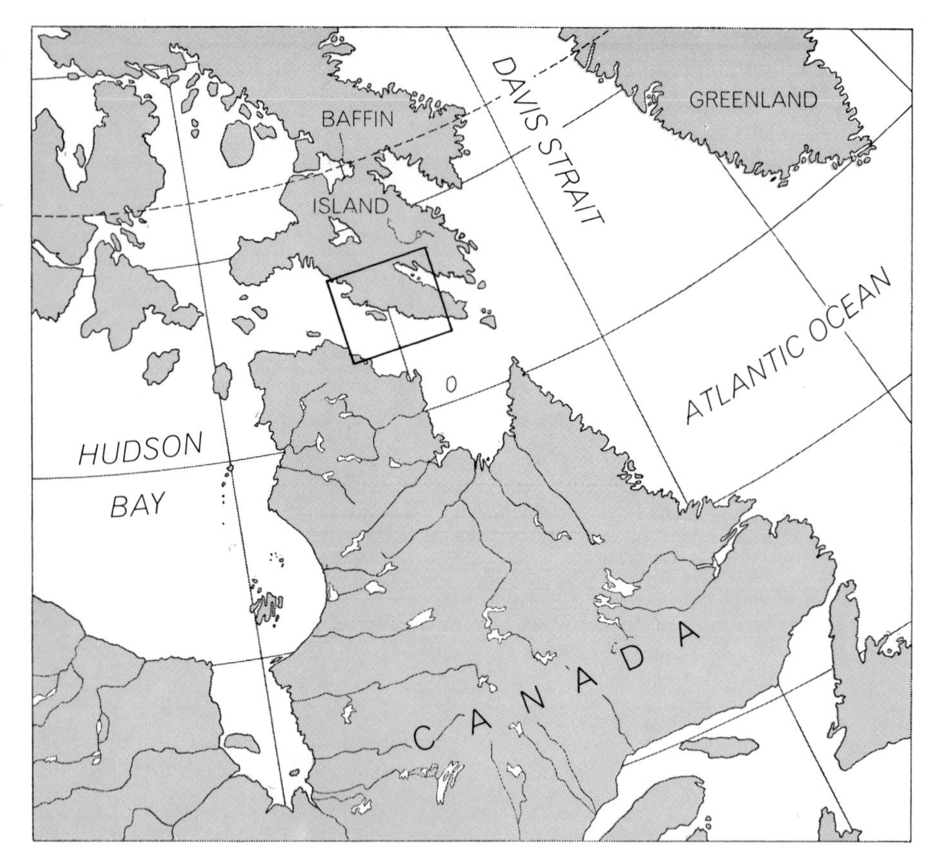

BAFFIN ISLAND extends for nearly 1,000 miles in the area between the mouth of Hudson Bay and the Greenland coast. The villagers' hunting ranges lie within the black rectangle.

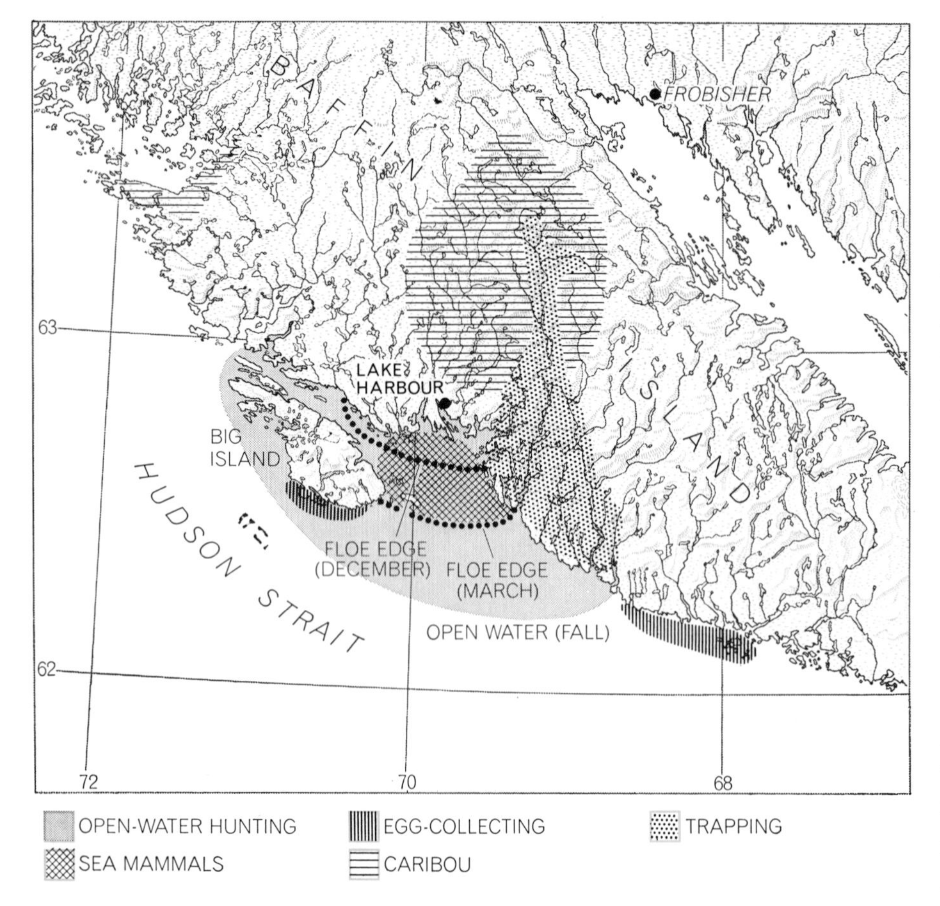

HUNTING RANGES of one Eskimo village in southern Baffin Island change with the season. The most productive months of the year are spent hunting in coastal waters. Trapping and caribou hunting, generally winter activities, carry the hunters into different areas.

tidal flats, leeward waters and protected valleys—are utilized. Traditionally the Eskimos of southern Baffin Island are mainly hunters of sea mammals: the small common (or ring) seal, the much larger bearded seal and on occasion the beluga whale and the walrus.

Survival in such a harsh environment has two primary requirements. The first is an adequate caloric intake in terms of food and the second is maintenance of a suitable microclimate in terms of shelter and clothing. In the village where I worked the Eskimos met these requirements by hunting and trapping and by buying imported foods and materials. They also bought ammunition for their guns and gasoline for two marine engines and two snowmobiles, transportation aids that increased their hunting efficiency. The money for such purchases was obtained by the sale of skins and furs (products of the hunt) and of stone and ivory carvings (products of artistic skill and of the Eskimos' recognition of consumer preferences). Because government "social assistance" and some work for wages were available, certain individuals had an occasional source of additional cash.

The daily maintenance of life in the village called for an initial input of human energy in the pursuit of game, in the mining of soapstone and in the manufacture of handicrafts. The expenditures of human energy for the 54-week period were 12.8 million kilocalories. They were augmented by expenditures of imported energy: 10,900 rounds of ammunition and 885 gallons of gasoline. The result was the acquisition of 12.8 million kilocalories of edible food from the land and seven million kilocalories of viscera that were used as dog food. To this was added 7.5 million kilocalories of purchased food. By eating the game and the purchased food the hunters and their dependents were able to achieve a potential caloric input of 3,000 kilocalories per day, which was enough to sustain a level of activity well above the maintenance level. The general pattern of energy flow into, through and out of the village is shown in the illustration on pages 58 and 59.

During my stay the population of the village varied between 26 and 29. The people lived in four separate dwellings. One, a wood house, had been built from prefabricated materials supplied by the government. The other three were the traditional *quagmaq:* a low wood-frame tent some 20 feet long, 15 feet wide and seven feet high. These structures were covered with canvas, old

BEARDED SEAL (*left*) is a relatively uncommon and welcome kill. It weighs more than 400 pounds, compared with the common seal's 80 pounds, and its skin is a favorite material. Flat-bottomed rowboat (*rear*) is used to retrieve seals killed at the floe-ice edge.

mailbags and animal skins and were insulated with a 10-inch layer of dry shrubs. Inside they were lined with pages from mail-order catalogues and decorated with a fantastic array of trinkets and other objects. The rear eight feet of each *quagmaq* was occupied by a large sleeping platform, leaving some 180 square feet for household activities during waking hours. The wood house (Household II) was occupied by six people comprising a single family unit. One *quagmaq* (Household I) was occupied by nine people: a widower, his son, three daughters, a son-in-law and three grandchildren.

The *quagmaq* was heated in the traditional manner with stone lamps that burned seal oil. The occupants of the wood house heated it with a kerosene stove. In a period when the highest outdoor temperature during the day was −30 degrees, I measured the consumption of fuel and recorded the indoor temperatures of the wood house and the *quagmaq*. The *quagmaq* was heated by three stone lamps, two at each side of the sleeping platform and one near the entrance. In a 24-hour period the three lamps burned some 250 ounces (slightly less than two imperial gallons) of seal oil. The fat from a 100-pound seal shot in midwinter yields approximately 640 ounces of oil, which is about a 60-hour supply at this rate of consumption. The interior temperature of the *quagmaq* never rose above 68 degrees. The average was around 56 degrees, with troughs in the low 30's because the lamps were not tended through the night.

In the wood house the kerosene stove burned a little less than three imperial gallons every 24 hours. This represented a daily expenditure of about $1 for fuel oil during the winter months. (Since 1969 this cost has been fully subsidized by the government.) The interior temperature of the house sometimes reached 80 degrees, and the nightly lows were seldom below 70. One result of the difference between the indoor and outdoor temperature—frequently more than 100 degrees—was that the members of Household II complained that they were uncomfortably warm, particularly when they were carving or doing some other kind of moderately strenuous work.

Before the Eskimos of southern Baffin Island had acquired outboard motors, snowmobiles and a reliable supply of fuel they shifted their settlements with the seasons. Fall campsites were the most stable element in the settlement pattern and were the location for the *quagmaq* shelters. During the rest of the year the size and location of the camps depended on which of the food resources were being exploited.

The movements of settlements to resources served to minimize the distance a hunter needed to travel in a day. Thus little energy was wasted traversing unproductive terrain, and good hunting weather could be exploited immediately. This is no longer the practice. Long trips by the hunters are common, but seasonal movements involving an entire household are rare. Therefore the location of the village never shifts. The four dwellings of the village I studied are more or less occupied throughout the year. Tents are still used in summer, but they are set up within sight of the winter houses. There is a major move each August, when the villagers set up camp at the trading post, a day's journey from the village. There they await the coming of the annual medical and supply ships and also take advantage of any wage labor that may be available.

The hunters' ability to get to the right place at the right time is ensured by a large whaleboat with a small inboard engine, a 22-foot freight canoe driven by a 20-horsepower outboard motor, and the two snowmobiles. In addition the villagers have several large sledges and keep 34 sled dogs. The impact of motorized transport (particularly of the marine engines) on the stability of year-round residence and on the increase in hunting productivity is evident in the remark of an older man: "As my son gets motors for the boats, we are always living here. As my son always gets animals we are no longer hungry. Do you know what I mean?"

The threat of hunger is a frequent theme in village conversation, but the oral tradition that serves as history gives little evidence of constant privation for the population of southern Baffin Island. Although older men and women tell stories of hard times, the fear of starvation did not generate the kind of social response known elsewhere. In the more hostile parts of the Arctic female infanticide was common well into the first quarter of this century. The existence of the practice is supported by statistical data compiled by Edward M. Weyer, Jr., in 1932. For example, among the Netsilik Eskimos the ratio of females to males in the population younger than 19 was 48 per 100; in the Barren Grounds area the ratio was 46 per 100.

Census data from Baffin Island in 1912 indicate that female infanticide was not common along the southern coast. In that year a missionary recorded the population for the region; the total was some 400 Eskimos. Among those younger than 19 there were 89 females per 100 males. Among those 19 or older the ratio was 127 females per 100 males; hunters often had short lives. The vital statistics that have been kept since 1927 by the Royal Canadian Mounted Police support the impression that death by starvation was a rare occurrence on southern Baffin Island. On the other hand, hunting accidents were the cause of 15 percent of the deaths. The causes of trouble or death most usually cited by the villagers were peculiarities of the weather, ice conditions and mishaps of the hunt. Starvation was commonplace in the dog population, but for human groups disasters were local. A 75-year-old resident of the southern coast was able to recall only one year when severe hunger affected a large segment of the population.

A major factor in reducing the possibility of hunger is the Eskimos' increasing access to imported goods. Although store foods are obviously of prime importance in this respect, energy in the form of gasoline for fuel is also significant. The snowmobiles in the village are owned by an individual in each household, but all the hunters help to buy the gasoline needed to run these machines and the marine engines. The two snowmobiles consumed about twice as much fuel as the two boat motors. A snowmobile pulling a loaded sled can run for about 35 minutes on a gallon of gasoline. A trip from the village to the

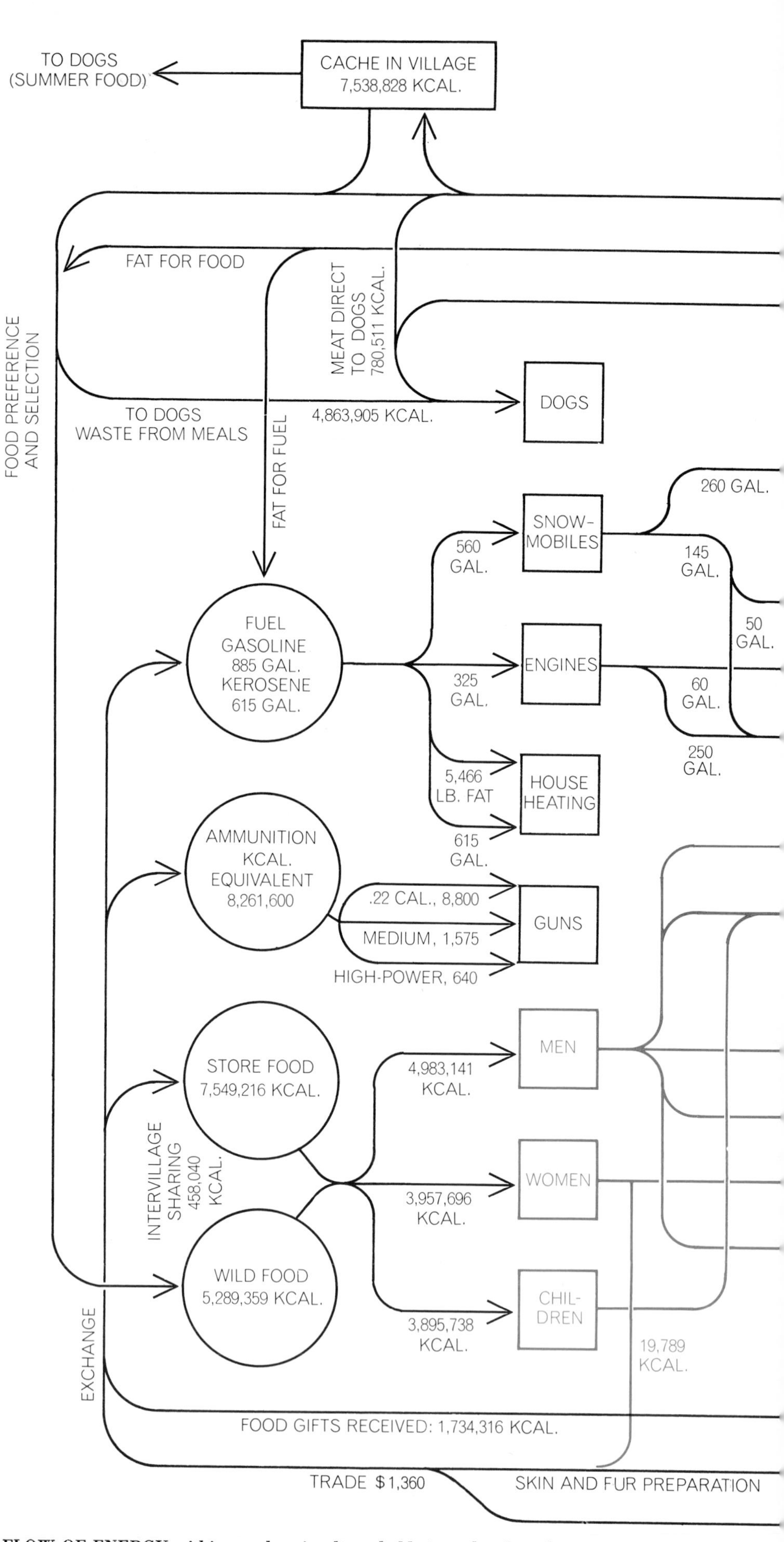

FLOW OF ENERGY within two hunting households is outlined in this diagram. The inputs and yields were recorded by the author in kilocalories and other units during his 13-month residence in an Eskimo hunting village. The input of imported energy in the form of fuel and ammunition, along with the input of native game and imported foodstuffs (*far*

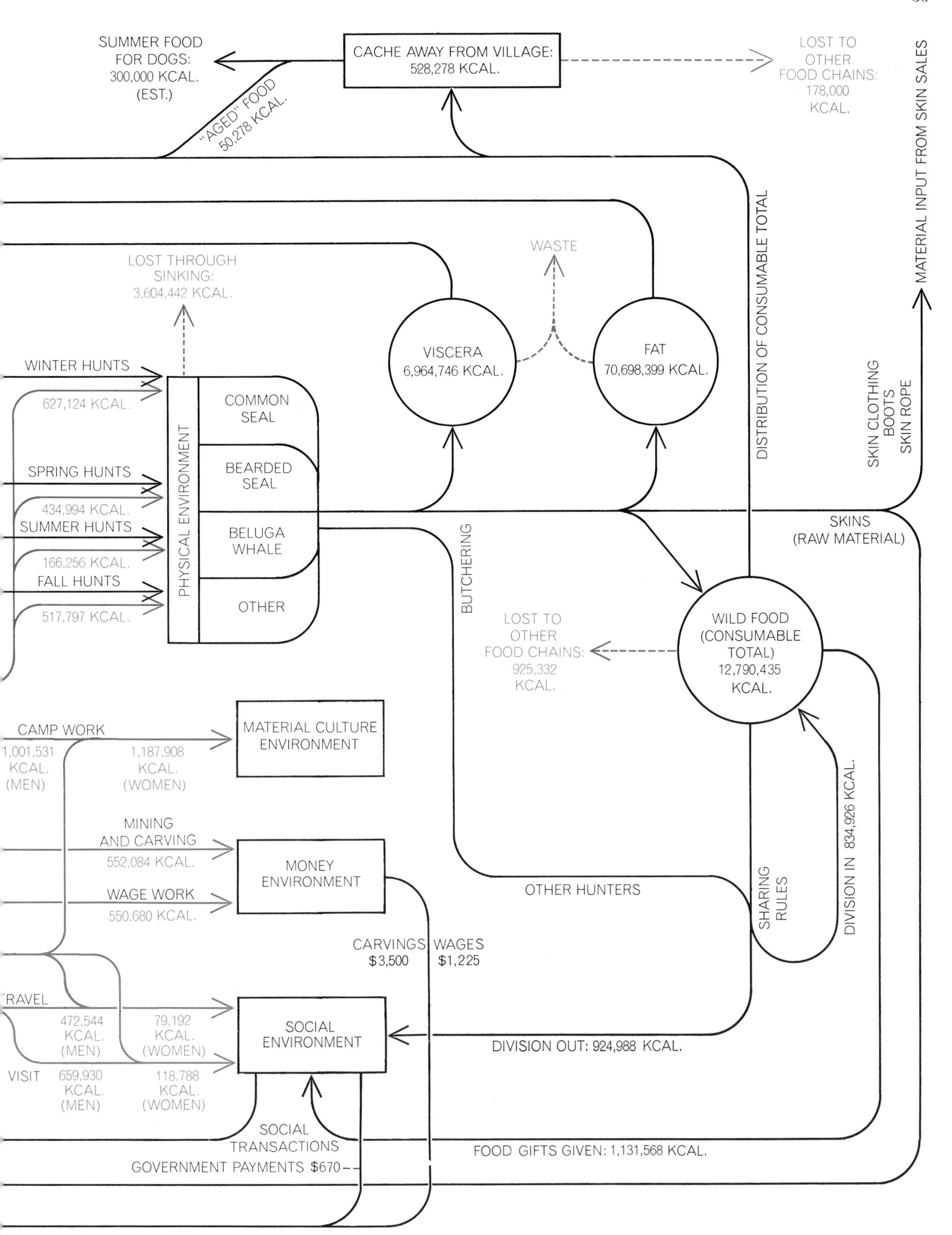

left), enabled the four hunters and their kin (*left, color*) to heat their dwellings and power their machines (*left, black*), and also to join in many seasonal activities (*colored arrows*) that utilized various parts of the environment in the manner indicated (*right*). **The end results of these combined inputs of energy are shown as a series of yields and losses from waste and other causes (*far right*). The net yields then feed back through various channels (*lines at borders of diagram*) to reach the starting point again as inputs.**

edge of the floe ice 10 miles away took 55 minutes in each direction and cost nearly $3.

Variations in fuel purchases are not necessarily correlated with variations in hunting yield or cash income. Debt can be used to overcome fluctuations in income, and the desire to visit distant relatives may be as important a consideration as the need to hunt. The largest monthly gasoline purchase was made in September, 1967. Wages paid for construction work were used to buy a combined total of 170 gallons of gasoline. This fuel was utilized for the intensive hunting of sea mammals in order to make up for a summer when the Eskimos had been earning wages instead of hunting.

Hunting was a year-long village occupation in spite of considerable seasonal variation in the kind and amount of game available. An analysis of the species represented in the Eskimos' annual kill confirms the predominance of sea mammals. The common seal (with an average weight of 80 pounds) provided nearly two-thirds of the villagers' game calories. When one adds bearded seals (with an average weight of more than 400 pounds) and occasional beluga whales, the sea mammals' contribution was more than 83 percent of the annual total. Caribou accounted for a little more than 4 percent of the total, and all the other land mammals together came to less than 1 percent. Indeed, the contribution of eider ducks and duck eggs to the villagers' diet (some 7 percent) was larger than that of all land mammals combined. The harvest of birds, fish, clams and small land mammals may not contribute significantly to the total number of calories, but it does provide diversity in hunting activities and in diet.

The common seal is hunted throughout the year and is the basic source of food for both the Eskimos and their dogs. From January through March sea mammals are hunted first at breathing holes in the sea ice and then at the boundary between open water and the landfast ice. The intensity of sea-mammal hunting during the winter months varies according to the amount of food left over from the fall hunt and the alternative prospects for trapping foxes and hunting caribou.

In winter some variety in the food supply is provided by the hunting of sea birds and small land mammals, but for the most part seal meat remains the basic item in the diet. In April hunting along the edge of the floe ice is intensified, and the canoe is hauled to the open water beyond the floe ice for hunting the bearded seals. In May and June hunting along the edge of the floe ice (on foot or by canoe) continues; the quarry is the seal and the beluga whale. In late spring seals are also stalked as they bask on top of the ice. By the middle of July open-water hunting is the most common activity, although much of the potential harvest is lost because the animals sink when they are shot. In 1967 and 1968 the villagers lost five whales, five bearded seals and 47 common seals.

The sinking of marine mammals serves to illustrate the interplay of physical, biological and technological factors the hunter must contend with. In late spring the seal begins to fast and therefore loses fat. At the same time the melting of snow and sea ice reduces the salinity of the surface waters. These interacting factors reduce the buoyancy of the seal, and a killed seal is likely to sink unless it is immediately secured with a hand-thrown harpoon. The high-powered rifle separates the hunter from his prey; it may increase the frequency of kill but it does not increase the frequency of harvest. In 30 hours of continuous hunting on July 20 and 21 only five out of the 13 seals killed were actually harvested.

From May through July the hunts are usually successful, even with sinkage losses as high as 60 percent. It is at this time of the year that a large amount of meat goes into dog food for the summer and early fall. Meat is also cached in areas the Eskimos expect to visit when they are trapping the following winter. The caches are deliberately only partly secured with rocks; their purpose is to bait areas of potential trapping. In May the variety of foods begins to increase. Seabirds, ptarmigans, geese, fish, clams and duck eggs are taken in large numbers and become the most important component of the food input. Only an occasional seal or the edible skin of a beluga whale is carried home to eat. The great variety of small game is consumed within a few days. With the exception of duck eggs, about half of which are cached until after Christmas, none of these foods is stored.

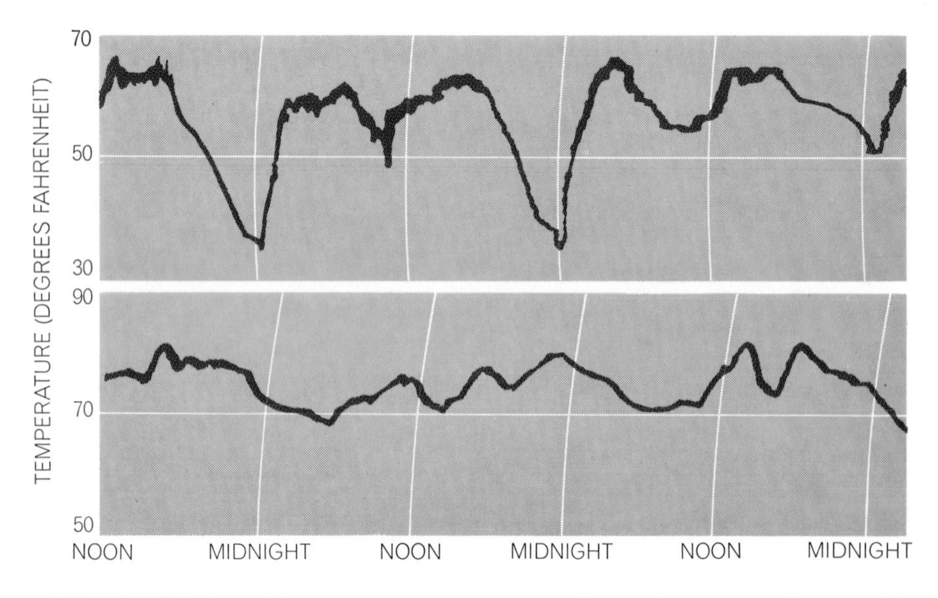

CONTRASTING METHODS of heating maintained different house microclimates with the consumption of different amounts of fuel. In Household I (*top*), the more traditional one, the use of three lamps that burned seal oil produced an average temperature of 58 degrees F. A kerosene stove (*bottom*) in the more modern Household II kept the average closer to 75 degrees. As a result Household II used three gallons a day to the other's two.

In September the sea mammals again become the primary objective. Open-water hunting continues until early November, when the sea begins to freeze. Just before freeze-up the beluga whales pass close to the village and are hunted from the shore with the aid of rifles and harpoons. The success of the fall whale-hunting is the key to the villagers' evaluation of the adequacy of their winter provisions.

As the sea ice thickens and extends, hunting seals at their breathing holes in the bays becomes the most common activity. The unfavorable interaction of physical, biological and technological factors that affects open-water seal-hunting in spring and summer is reversed where breathing-hole hunting is concerned. As the sea begins to freeze, some seals migrate away from the land in order to stay in open water. Others remain closer to the shore, using their claws and teeth to maintain a cone-shaped hole through the ice. The seal's breathing is now confined to specific points the hunter can easily find on the surface of the ice. The hunting technique calls for locating the breathing holes and distributing the available men to maximize the chance of a kill. The hunting skill calls for the patience to wait motionless for periods of as much as two hours and to depend on hearing rather than sight. By mid-December the new ice is covered by a deep layer of drifting snow, and the breathing holes become harder to locate. The Eskimos then move out to the edge of the floe ice and the seasonal cycle begins anew.

The analysis of hunting success on a month-to-month basis shows great variability in the total caloric input and in each member's contribution to the total [*see illustration on next page*]. The peak hunting months for the two households were June (some 2.5 million kilocalories), October (more than three million) and November (2.9 million). Stockpiling provides the motivation for the big October and November kills. Game taken in these months will remain frozen and unspoiled through the winter and will help to feed the hunters and their dependents in February, March and April.

In the fall days grow short and winds often restrict the choice of hunting areas. Daylight hours are utilized to the full, and the evening darkness is filled with the sound of the hunters struggling to get their catch across the difficult terrain of the tidal flats. In this period almost all the food is brought back to the village; it is stored in a small meat house, on elevated platforms, under the hulls of old boats and on top of the wood house. A few of the seals shot in early fall are cached on the land in order for the meat to "age." These carcasses are retrieved in the spring, and the meat is considered one of the more flavorsome food inputs.

The large kill in June results from the fact that the daylight hours are at a maximum and that there is a much greater choice of resources. If weather conditions hamper open-water hunting, basking seals can be pursued. Under conditions of severe wind or poor ice, spearfishing for arctic char is possible and duck eggs can be collected. Summer hunts last for three or four days; the hunters sleep during the few hours of least light or during brief pauses in the hunt. Game killed in June will thaw in the summer months, so that almost all the two million kilocalories of sea mammals harvested that month is destined for the dogs.

Compared with the high caloric inputs of the spring and fall, winter hunting is much less productive in terms of total harvest. For example, February, 1967, was only a fair hunting month for Household I and a very bad month for Household II. The combined kill (more than 70 percent common seals) provided only 166,500 kilocalories of food; more than 90 percent of the total was taken by Household I. A month of low food input does not, however, mean hardship. Such a February is an example of the important role storage plays in the villagers' management of energy resources. The fall hunt had provided enough food for the winter, and as a result in February the villagers did more visiting than hunting. Visiting is therefore one mechanism that takes hunters out of the productive sector of the economy and creates a better balance between energy availability and energy need. The same pattern holds true throughout the winter months, so that the 500,000 calories that was harvested from February through April was as much a caloric expression of leisure as it was of poorer hunting conditions. The differential hunting success of individuals or of households in the month of February did not greatly influence energy distribution within the social unit. Although one hunter may have more skins to trade, food is stored in bulk and is generally available to all.

The records show that although caribou are hunted only occasionally, the animals are then present in substantial numbers. In the lean February of 1967 caribou made up some 7 percent of the kill, providing about 11,000 kilocalories of food. The following month caribou comprised more than 37 percent of the kill, amounting to a total of 90,000 kilo-

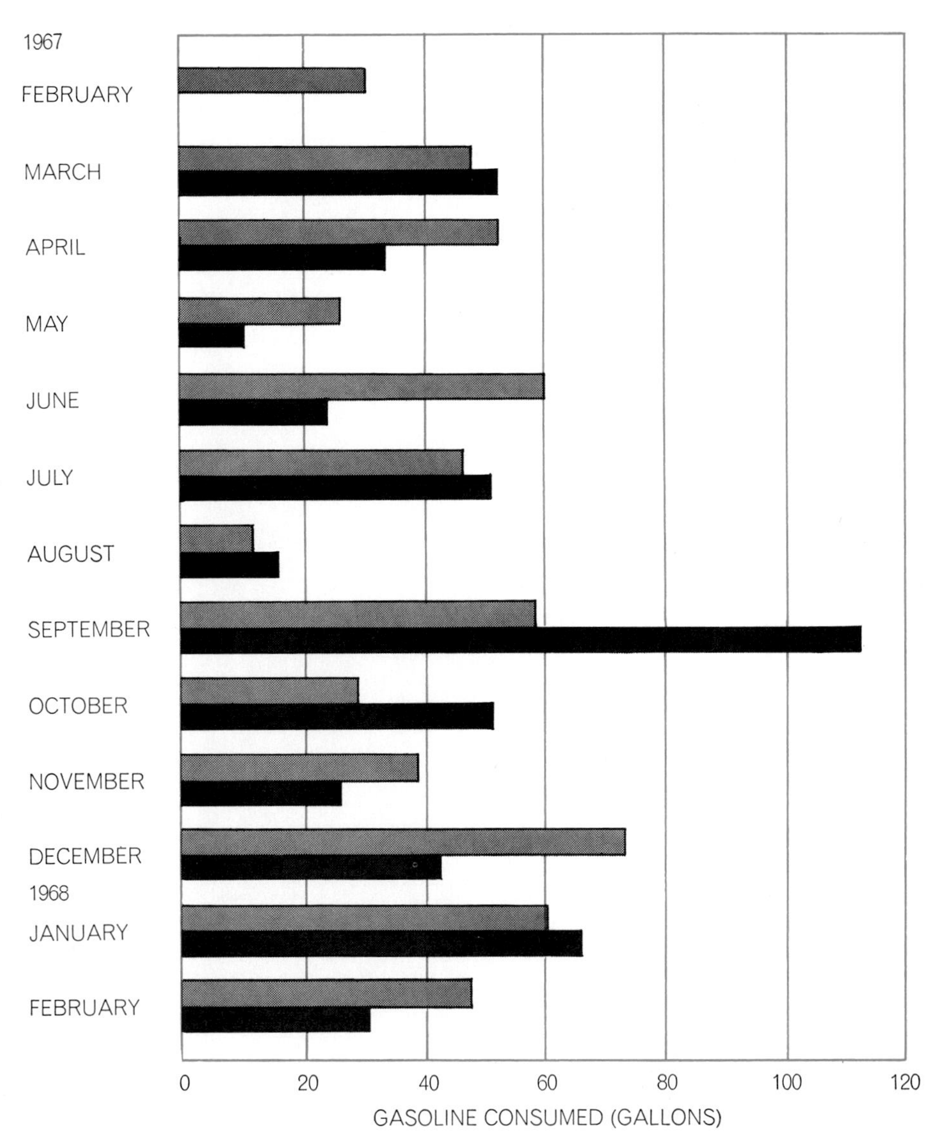

GASOLINE CONSUMPTION by two Eskimo households is shown over a 12-month period. The fuel was used to power the two snowmobiles and the marine engines that greatly increased the villagers' hunting efficiency. Purchases by the three hunters of Household I are shown in black and those by the single hunter of Household II are in gray. Gasoline is second only to imported food among the exotic energy inputs to the Eskimo hunting society.

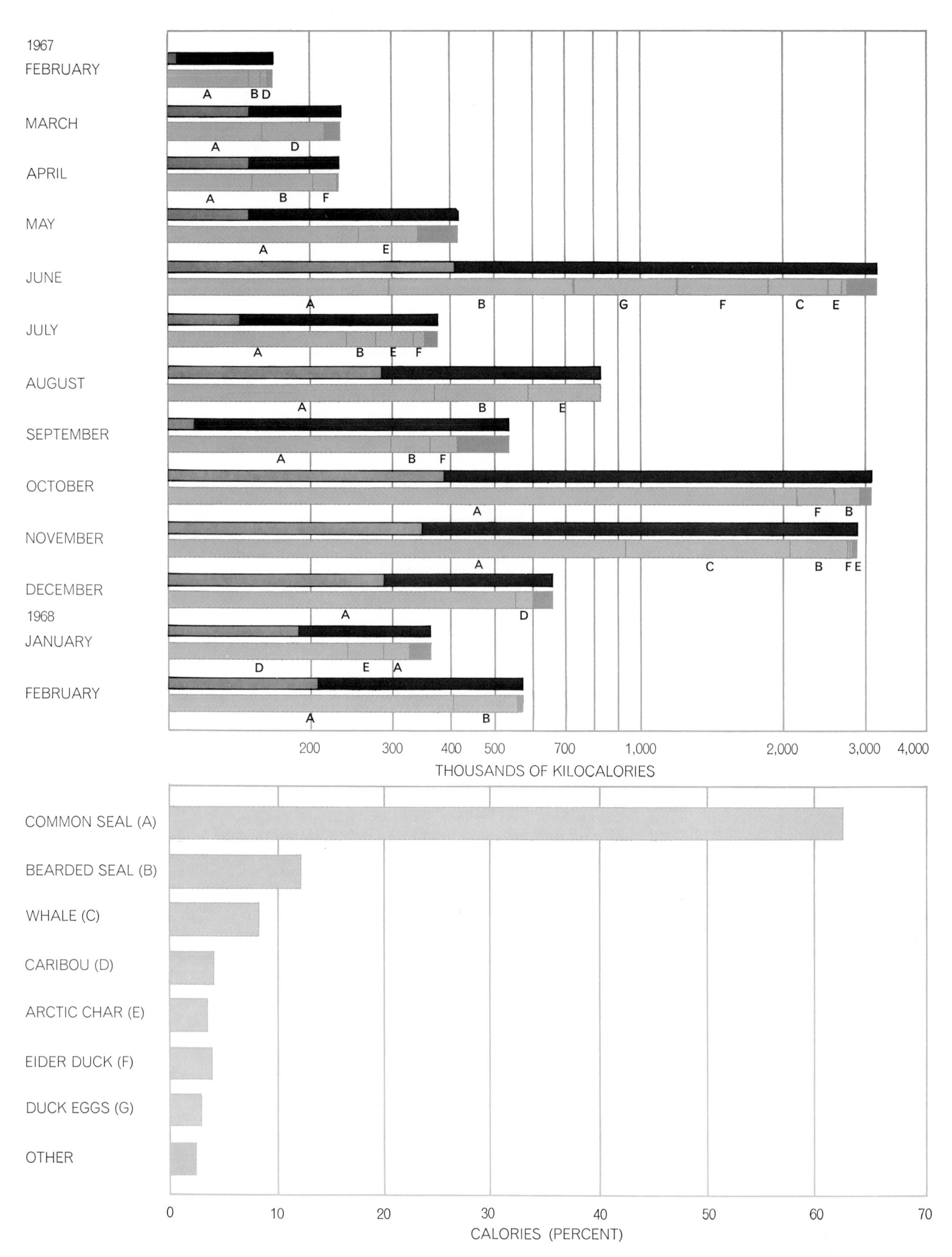

HUNTERS' BAG varies considerably from month to month as a result of chance and preference and also because of seasonal fluctuations. The top graph shows the wild foods acquired by Household I (*black*) and Household II (*gray*) in the course of 13 months. A fish known as arctic char, birds such as murres, geese and ducks, duck eggs and even berries add variety to the Eskimo diet from April through October, while caribou contribute to the smaller game bag of winter months. The 13-month totals, however, show that sea mammals (the common seal in particular) provide most of the Eskimo households' consumable kilocalories (*bottom graph*).

calories. After that caribou were almost absent from the villagers' diet until January, 1968, when they furnished nearly 70 percent of the kill for a yield of 245,000 kilocalories.

Today no Eskimo community depends exclusively on hunting for its food. Each day the adults of the village consumed an average of half a pound of imported wheat flour in the form of a bread called bannock, a pan-baked mixture of flour, lard, salt and water. Bannock has long been an Eskimo staple; it is eaten in the largest quantities where hunting has fallen off the most.

In addition to this basic breadstuff the villagers consumed imported sugar, biscuits, candy and soft drinks, and they fed nonnursing infants and young children a kind of reconstituted milk. A daily ration consisted of 48 ounces of water containing 1.2 ounces of dry whole milk and 1.7 ounces of sugar.

I kept a 13-month record of the kind and amount of imported foods bought by the two households. During this period the more traditional household bought store foods totaling almost 3.3 million kilocalories and the more modern household store foods totaling 3.75 million kilocalories. The purchases provided 531,000 kilocalories of store food per adult in Household I and 477,300 per adult in Household II. The larger number of adults in Household I is reflected in the size of its flour purchase, which made up 53 percent of the total, compared with 40 percent in Household II. The larger purchase of lard by Household II (23 percent compared with 10 percent) is a measure of preference, not consumption. Household I prefers to use the fat from whales for bannock; hence its smaller purchase [*see illustration on next page*]. The consumption of the third imported staple—sugar—was about the same in both households.

The quantities of store food that were bought from month to month showed substantial variations. In March, 1967, the store purchases of the two households rose above one million kilocalories because both households received a social-assistance payment. In February, 1968, social assistance was again given each household, and as before the money was used to buy more than the average amount of food. The rather high caloric input from store food in September, 1967, is attributable to money available from wage labor.

Store foods, unlike food from the land, are not stockpiled and they are not often shared. Except for the staples and tea, tobacco and candy, there is no strong desire for non-Eskimo foods. When vegetables are bought, it is usually by mistake. Canned meats, although occasionally eaten, are not recognized as "real" food. The villagers like fruit, but it is never bought in large quantity. Jam, peanut butter, honey, molasses, oatmeal and crackers all find their way into the two households. They are consumed almost immediately.

The data on food input support the general findings from other areas that show the Eskimo diet to be high in protein. At least in this Eskimo group, even though imported carbohydrates were readily available and there was money enough to buy imports almost ad libitum, the balance was in favor of protein.

Over the 13-month period the villagers acquired 44 percent of their calories in the form of protein, 33 percent in the form of carbohydrate and 23 percent in fat. Almost all the protein (93 percent) came from game; 96 percent of the carbohydrate was store food. The figures suggest how nutritional problems can arise when hunting declines. As store-food calories take the place of calories from the hunt, the change frequently involves increased flour consumption and consequently a greater intake of carbohydrate. This was the case in Household II during September, 1967, a period when the family worked for wages. The caloric input remained at 2,700 kilocalories per person per day, but 62 percent of the calories were carbohydrate and only 9 percent were protein.

A framework of social controls surrounds all the activities of the village, directing and mediating the flow of energy in the community. For example, even though all the inhabitants are ostensibly related (either by real kinship or by assigned ties), the community is actually divided into two social groups, each operating with a high degree of economic and social independence. Food is constantly shared within each social group, and the boundary between groups is ignored when a large animal is killed.

Village-wide meals serve to divert a successful hunter's caloric acquisitions for the benefit of a group larger than his own household. The meal that follows the arrival of hunters with a freshly killed seal is the most frequent and the most important of these events. It is called *alopaya,* a term that refers to using one's hands to scoop fresh blood from the open seal carcass. The invitation to participate is shouted by one of the children, and all the villagers gather, the men in one group and the women in another, to eat until they are full. The parts of the seal are apportioned according to the eater's sex. The men start by eating a piece of the liver and the women a piece of the heart. The meat from the front flippers and the first third of the vertebrae and the ribs goes to the women. The men eat from the remaining parts of the seal. This meal, like almost all other Eskimo meals, does not come at any specific time of the day. People eat when they are hungry or, in the case of village-wide meals, when the hunters return. If anything remains at the end of an *alopaya* meal, the leftovers are divided equally among all the families and can be eaten by either sex.

Whaleboats and freight canoes began to replace the kayak for water transportation in southern Baffin Island some 30 years ago; by the end of the 1950's outboard motors had been substituted for oars and paddles. The latter change, which made possible more efficient open-water hunting, coincided with a high market price for sealskins. In the early 1960's a single skin might bring as much as $30, and the value of the annual village catch was between $3,000 and $4,000. The good sealskin market enabled the villagers to buy their first snowmobile in the fall of 1963. A decline in skin prices that began in 1964 has since been offset to some extent by the growth of handicraft sales and by the availability of work for wages.

As a result of these new economic inputs a new kind of material flow is now observable. It consists of the movement of secondhand non-Eskimo goods. The flow is channeled through a network of kinship ties between individuals with an income higher than average and those with an income lower. By 1971 anyone who wanted a snowmobile or some other item of factory-made equipment could utilize this network and get what he wanted through a combination of salvage, gift and purchase.

For the individual the exploitation of economic alternatives and the pattern of activity vary according to taste and life-style. Although life-style is in large degree dictated by age, the effect within the village is integrative rather than disruptive. A son's snowmobile gives the father the advantage of a quick ride to the edge of the floe ice, and the son can rely on the father's dogs to tow a broken machine or to help pull heavy loads over rough terrain. The integration extends to areas beyond the hunt. At home it is not unusual to find the father making sealskin rope or repairing a sledge while the son carves soapstone. Neither considers the other's work either radical or impossibly old-fashioned.

Social controls also affect the expendi-

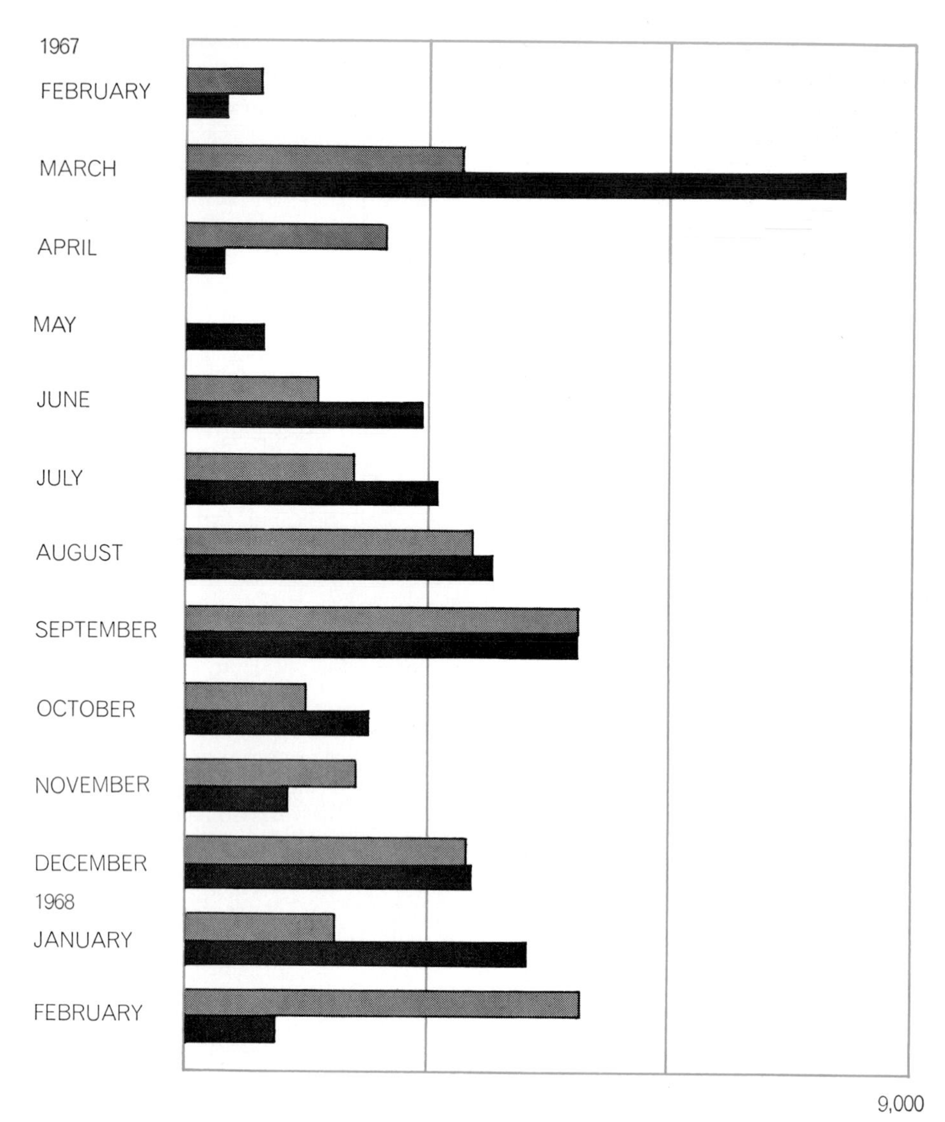

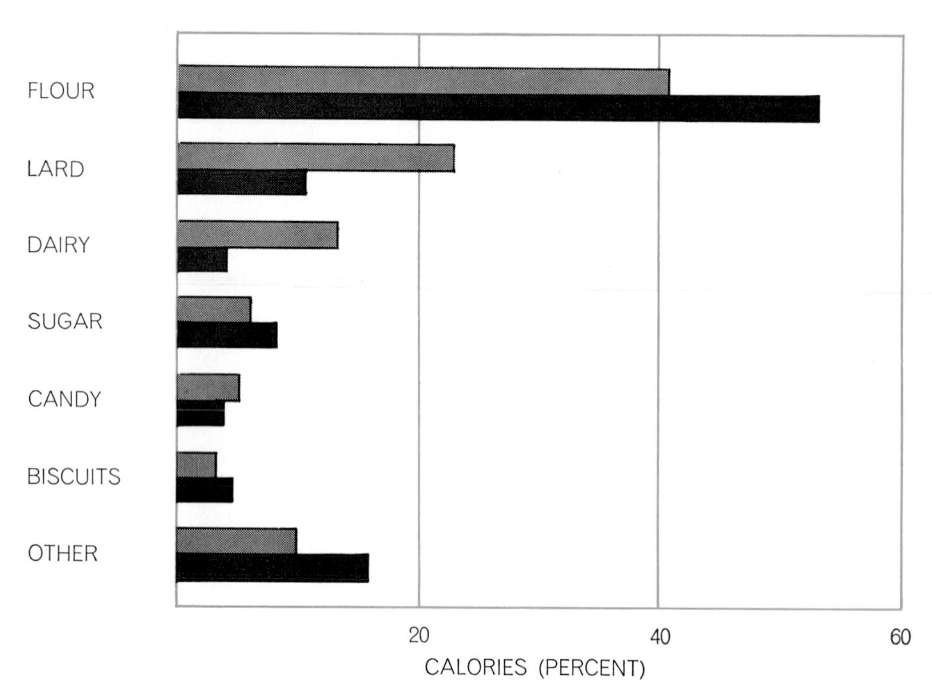

PURCHASES OF IMPORTED FOOD also show large monthly variations. The top graph shows the kilocalorie values of staples such as flour, lard and sugar and of lesser items such as powdered milk, biscuits, soft drinks and candy acquired over 13 months by Household I (*black*) and Household II (*gray*). Most of the flour and lard went to make a kind of bread called bannock. The 13-month totals (*bottom graph*) show how the two households differed in the percentage of all store-food purchases that each allotted to flour and to lard.

ture of personal energy within the household. The losers are the teen-age girls. Among the men of the village there is little emphasis on authority structure or leadership; decision-making is left to the individual. Choices for the most part converge, so that joint efforts are a matter of course. Among the women, however, authority structure is emphasized. A girl is subordinate not only to the older women but also to her male relatives. One can make the general statement that those who because of sex, age and kinship ties are most subject to the demands of others expend a disproportionate amount of energy in household and village chores.

One series of social controls has been radically altered by the introduction of non-Eskimo technology, energy and world view. These controls are the beliefs and rituals relating the world of nature to the world of thought. In the traditional Eskimo society the two worlds were closely related. All living organisms, for example, were believed to have a soul. In his ritual the Eskimo recognized the fragility of the Arctic ecosystem and sought to foster friendly relations with the same animals he hunted for food. Obviously the friendship could not be a worldly one, but it did exist in the realm of the spirit. A measure of the strength of this belief was the great care taken by the Eskimo never to invite unnecessary hardship by offending the soul of the animal he killed.

Today ritual control of the forces of nature and of the food supply has almost disappeared; technology is considered the mainspring of well-being. Prayers may still be said for good hunting and good traveling conditions, and the Sunday service may include an analysis of a hunting success and even a request for guidance in the hunts to follow. Hunting decisions may also be affected by dreams. None of these activities, however, has the regulatory powers of the intricate symbols and beliefs of earlier times. The hunters complain of a change in the seals' behavior. Nowadays, they say, only the young animals are curious and can be coaxed to come closer to the boat or the edge of the floe ice. The mutual trust between man and his food supply has evidently been lost in the report of the high-powered rifle and the rumble of the outboard motor.

What conclusions can we draw from the analysis of energy flows in a hunting society? The Eskimos do not differ from other hunters in that the processes surrounding the quest for food involve much more than a simple interplay of environment and technology. There are many times when a technological advance is fatal to an ecological balance; this was particularly evident in the near-extermination of the caribou herds west of Hudson Bay with the introduction of the rifle and the relaxation of traditional beliefs. In southern Baffin Island, however, there is not yet any evidence of a trend toward "overkill." At the same time that motorized transport has enhanced the ability to kill game, other social and economic factors have acted to reduce the amount of time available for hunting and have kept the kill within bounds. Snowmobiles give quick access to the edge of the floe ice; they also make it easy to visit distant kinsmen. A regular day of rest on Sunday has a religious function; it also contributes to the management of energy resources.

Do hunting societies have a long-term future? In the case of the Eskimo one can reply with a conditional yes. The universal pressure on resources makes continued exploitation of the Arctic a certainty, and Eskimos should be able to profit from these future ventures. Already it is possible to see three distinct groups emerging within Eskimo culture. One of them consists of the wage earners in the larger communities. Another, which is just beginning to make its appearance, is an externally oriented group that seems destined to regulate and control the inputs from the outside world: the non-Eskimo energy flows and material flows that, as we have seen, now play a vital part in the hunters' lives.

Finally, there is a third group, small in numbers but vast in terms of territory, made up of the hunters who will continue the traditional Eskimo participation in the fragile, far-flung Arctic ecosystem. There will be linkages—exchanges of materials and probably of people—between the wage earner and the hunter. Those who choose to live off the land may appear to be the more traditional of the three groups, but their lives will be dynamic enough because the variables that define the hunting way of life are constantly changing. If a snowmobile is perceived to have greater utility than a dog sled, then the ownership of a snowmobile will become one of the criteria defining the traditional Eskimo hunter. With the outward-oriented Eskimos providing stability for the three-group system through their control of exotic inputs, the northern communities should be able to evolve further without developing disastrous strains. But the fundamental linkage—the relation between the hunter and the Arctic ecosystem—will remain the same.

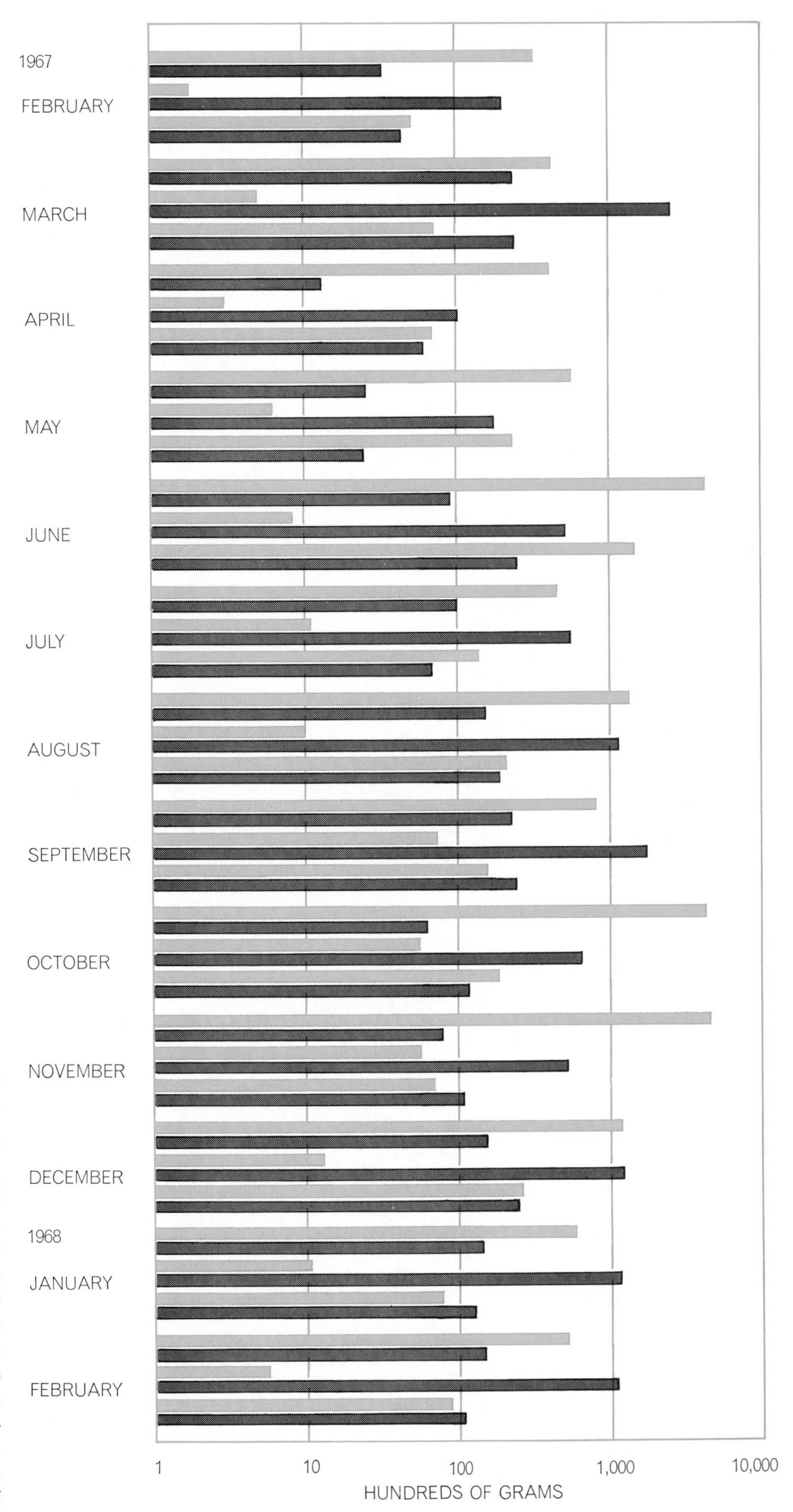

COMPOSITION OF DIET is presented for a 13-month period in terms of monthly acquisitions of protein (*top pair of bars*), carbohydrate (*middle pair of bars*) and fat (*bottom pair of bars*), measured in hundreds of grams. The colored bar of each pair indicates the number of grams acquired by hunting and gathering and the black bar indicates the number acquired in the form of store food. Protein outranked the others in total acquisitions: 2.1 million grams, compared with 1.1 million grams of carbohydrate and .7 million grams of fat.

6

The Flow of Energy in an Agricultural Society

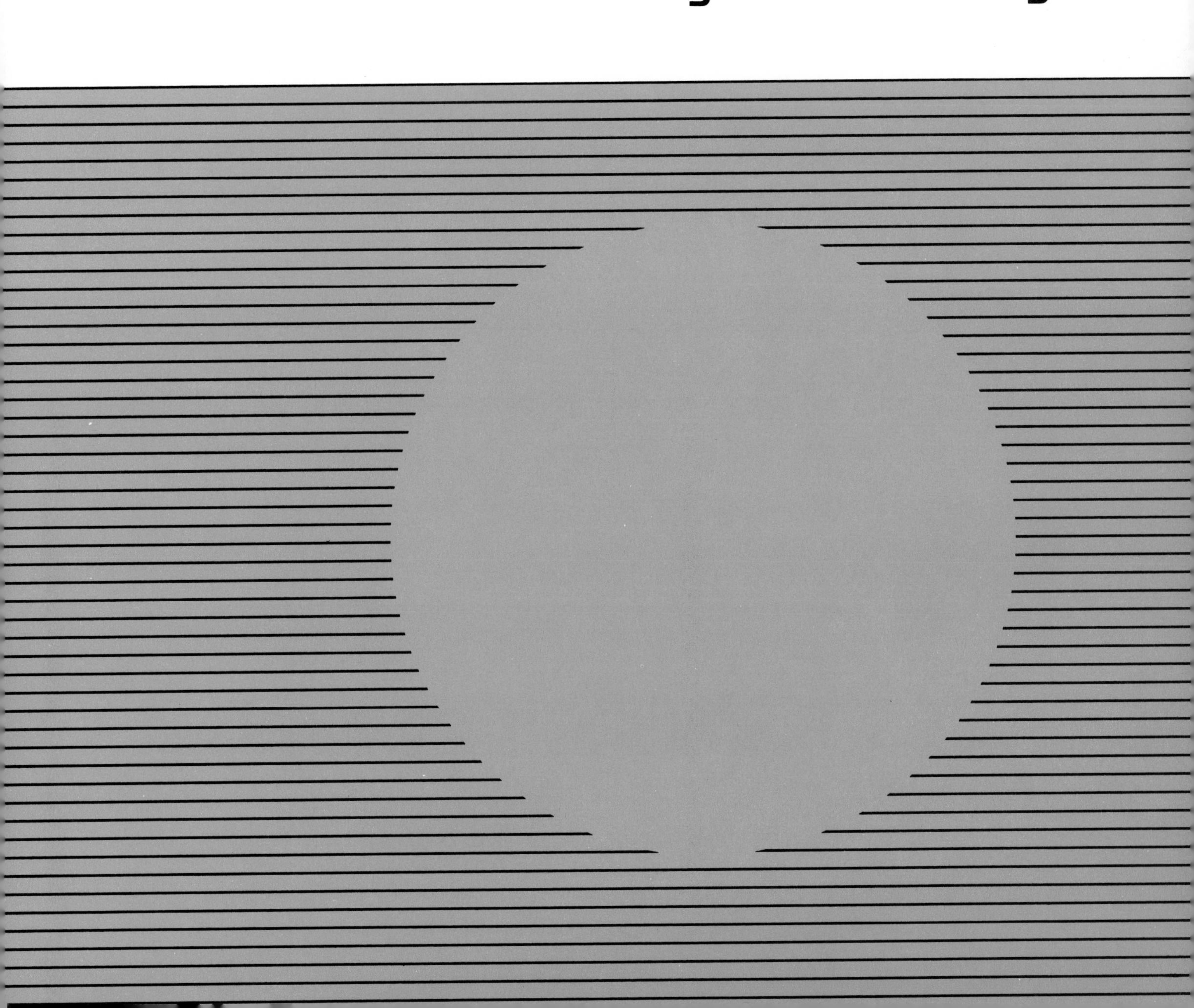

The Flow of Energy in an Agricultural Society

ROY A. RAPPAPORT

The invention of agriculture gave mankind a more abundant source of solar energy. The energetics of a primitive agricultural system are examined in New Guinea, with a moral for modern agriculturists

Raising crops and husbanding animals are man's most important means of exploiting the energy that is continuously stored in primary plant production. Man's manipulation through the practice of agriculture of this energy store and of the food chains it supports has enabled him to progress beyond the bare subsistence provided by hunting and gathering, and long ago placed human culture on the road leading to the complex social systems of today. Here we shall examine the flow of energy in an agricultural society that practices a mode of gardening known for millenniums, a mode likely to have been the first to enable pioneer farmers to exploit an almost unpopulated part of the world: the humid Tropics.

An examination of this type of gardening is particularly appropriate to the theme of this issue because the flow of energy within it is easy to trace; its practitioners have no power sources other than fire and their own muscle and have only the simplest tools. At the same time their kind of gardening and swine husbandry makes relatively light demands on the farmers in terms of energy inputs, provides for almost all their dietary needs and, if properly practiced, alters ecosystems less than other modes of agriculture of comparable productivity do. We shall compare this kind of farming with the ecologically more disruptive methods of modern agriculture, and we shall examine how the flow of energy and materials in agricultural systems affects the diversity and stability of ecosystems in general. We shall also consider the relation between social evolution, with its ever increasing demand for expenditures of energy, and ecological degradation.

The system of gardening is called "swiddening," and the place where I observed it is the tropical rain forest of New Guinea. The term comes from the Old Norse word for "singe." The method has often been applied in forest environments outside the Tropics, including the forests of medieval England where it got its name. It has many variants, but basic aspects of the procedure are much the same everywhere in temperate or tropical forests. A clearing is cut in the forest, the cuttings are usually burned (sometimes they are removed by hand or allowed to rot), a garden is planted and harvested and the clearing is then abandoned to the returning forest. On occasion the clearing is planted two or three times before it is abandoned, but a single planting is more typical.

The mature tropical rain forest is probably the most intricate, productive, efficient and stable ecosystem that has ever evolved. Men are able to use directly for food only a tiny fraction of the forest's biomass: its total store of living matter. The unmodified rain forest can support perhaps one human being per square mile. From the human viewpoint such an environment is seriously deficient, and swiddening provides a sophisticated and even elegant means of overcoming its deficiencies. With swiddening population densities comparable to those of industrialized countries are maintained with considerably less degradation of the environment resulting.

The swiddening farmers with whom we are specifically concerned are the Tsembaga, one of several local groups speaking the Maring language that live in the central highlands of the Australian Trust Territory of New Guinea some five degrees south of the Equator [*see top illustration on page 71*]. The Tsembaga occupy a territory on the southern side of the Simbai River valley in the Bismarck Range. From its lowest point along the riverbank, about 2,200 feet above sea level, their land rises in less than three miles to the mountain ridge at an altitude of 7,200 feet. In 1962 and 1963, when I was visiting the Tsembaga, their territory was 3.2 square miles and their population totaled 204.

Not all of the Tsembaga land is arable. The part that lies above 6,000 feet is usually blanketed by clouds and its vegetation consists of a moss forest unaltered by man. The next zone, between 6,000 and 5,000 feet, is cultivable on a marginal basis, but most of it has been left as unaltered mature rain forest. Below this zone, between 5,000 and 2,200 feet, is the main area of agricultural activity, although portions of the land are too rocky or too steep for gardening. Just over 1,000 acres here were occupied either by gardens or by fallow secondary-forest vegetation in 1962.

Forty-six acres—about .2 acre per person—had been newly planted that year. Since some gardens yield a harvest for two years or more, this practice means that as many as 90 to 100 acres of Tsembaga land are often in cultivation simultaneously. Conversely, at any one time at least 90 percent (and sometimes more) is lying fallow. If one adds to these 900-odd acres the 340 acres or so of marginally arable land in the zone between 5,000 and 6,000 feet that have never

GARDEN PLOTS that exemplify a very ancient method of farming lie scattered through the swath of second-growth forest in the aerial photograph on the opposite page. The agricultural area lies on the east bank of the Simbai River in the central highlands of New Guinea. Most gardens are located in the lower third of the area (*see detailed map on next page*). In any year 90 percent of the land lies fallow, slowly returning once again to forest.

been cultivated, the percentage of potentially arable land actually under cultivation becomes even smaller.

In terms of their entire territory the Tsembaga in 1962–1963 were maintaining a population density of 64 per square mile. In terms of all potentially arable land the density was 97 per square mile, and in terms of land that was then or ever had been under cultivation the density was 124 per square mile. Even this figure is below the carrying capacity of the Tsembaga territory; without altering the horticultural regime of keeping 90 percent of the land fallow the Tsembaga's 1,000 best acres might have supported a population of 200 or more per square mile.

Horticulture provides 99 percent of the everyday Tsembaga diet, but the unaltered forest beyond the gardens and the secondary forest that covers the fallow land also play an economic role. For example, the Tsembaga husbandry of pigs depends on feral boars that roam the forests. The Tsembaga castrate their own boars because they believe it makes for larger and more docile animals; the sows (which wander free during the day, returning home at dusk) must thus be impregnated by chance contact with feral boars. Feral pigs are also a source of some protein for the Tsembaga, as are the marsupials, snakes, lizards, birds and woodgrubs also found in the forest. Some forest animals and hundreds of species of forest plants provide the raw materials for tools, house construction, clothing, dyestuffs, cosmetics, medicines, ornaments, wealth objects and the supplies and paraphernalia of ritual. The greatest contribution of the forest, however, is providing a favorable setting for the Tsembaga gardens.

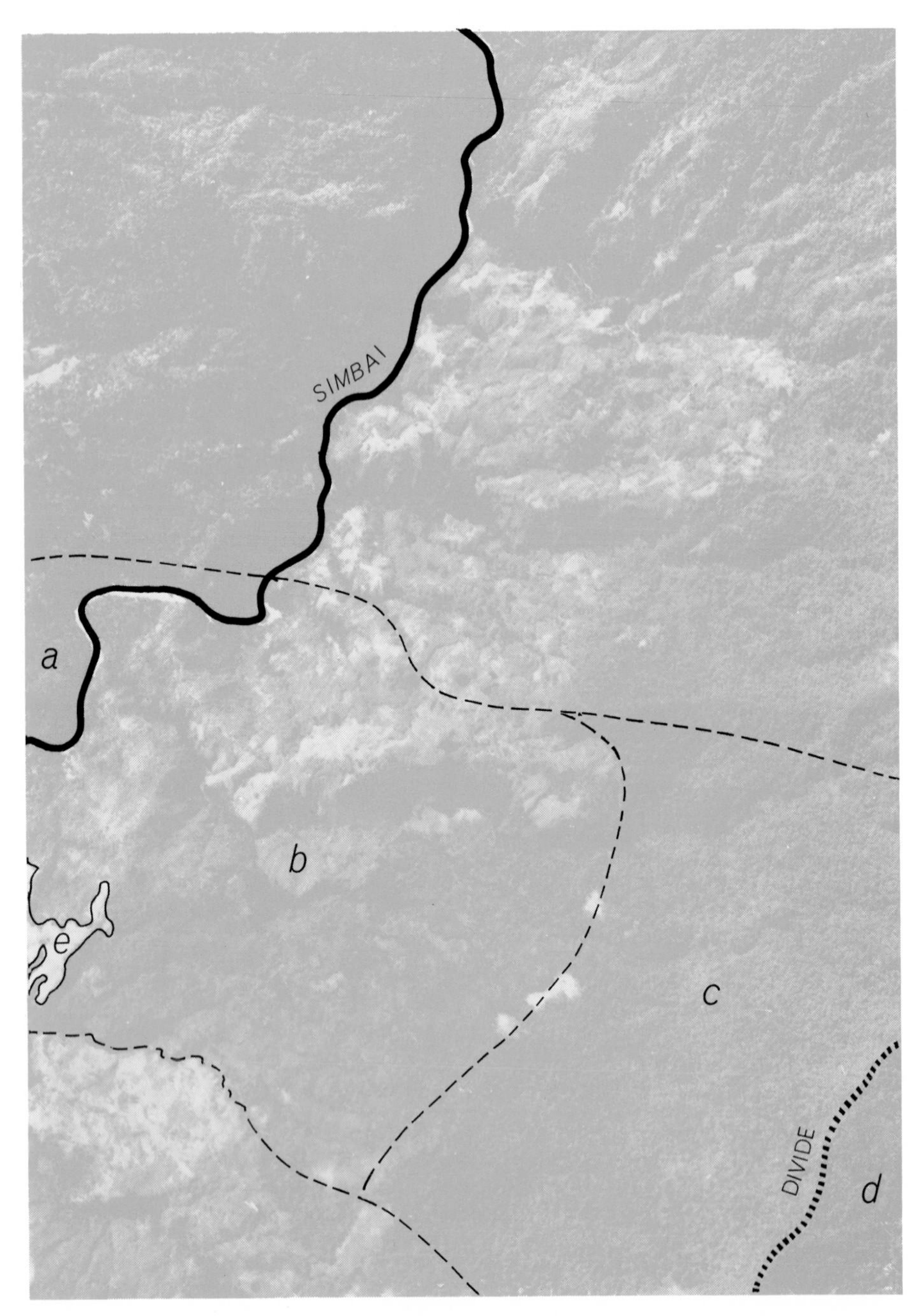

AREA OF STUDY has four major divisions. One forested zone *(a)* just north of the Simbai River is land as yet little used for gardens. The major agricultural zone is uphill from the river to the south *(b)*; it rises from about 2,300 feet above sea level at the river to about 5,200 feet at its boundary with a zone of largely virgin forest *(c)* farther uphill. The highest land, which is not farmed, lies in this zone and the adjacent one *(d)* on both sides of a mountain ridge some 7,200 feet above sea level. The light patch *(e)* is a stand of kunai grass that has sprung up in a cleared area and is resistant to the process of reforestation.

In 1963 I kept detailed records of the activities involved in transforming an 11,000-square-foot area of secondary forest into a garden. (These observations were supplemented by a wide range of measurements of similar activities of the same people and others at different sites.) Situated at an altitude of 4,200 feet, the land had been fallow for 20 to 25 years. On it, in addition to underbrush, were 117 trees with a trunk six inches or more in circumference, including a number that measured at least two feet. The canopy of leaves met some 30 feet overhead and had become dense enough to kill a good deal of the underbrush by its shade. In order to make estimates of energy input during various stages of the clearing work, I conducted time and motion studies in the field. These studies, in conjunction with the findings of E. H. Hipsley and Nancy Kirk of the Commonwealth of Australia Department of Health, provided a basis for my calculations. Hipsley and Miss Kirk, working with other New Guinea highlanders, the Chimbu, measured individual metabolic rates during the performance of everyday tasks. My estimates of crop yields are based on a daily weighing of the harvests from some 25 Tsembaga gardens over a period of almost a year. I have used various standard sources in calculating the energetic values of the produce.

In making a garden the Tsembaga prefer to clear secondary forest rather than primary forest because secondary growth is easier to cut and burn. Even in the secondary forest, clearing the underbrush is hard work, although it was made easier when machetes were introduced to the area in the 1950's. I found that the performance of men and women in clearing the brush was surprisingly uniform. Although in an hour some women clear little more than 200 square feet and some of the more robust men clear nearly 300 square feet, the larger men expend more energy per minute than the women. The energy input of each sex is approximately equal: some .65 kilocalorie per square foot, or 28,314 kilocalories per acre.

Once the underbrush is cut about two weeks are allowed to pass before the next step: clearing the trees. This is exclusively men's work. On this occasion most of the 117 trees in the garden were felled. Their branches were then lopped off and scattered over the piles of drying underbrush. Trees whose thick trunks would have been hard to burn and a nuisance to drag away were left standing, but most or all of their limbs were removed. The process of tree-clearing is far less strenuous than clearing underbrush: the energy investment is about .26 kilocalorie per square foot, or 11,325 kilocalories per acre.

The next step is to make a fence to keep out pigs, both feral and domestic. The trunks of the felled trees are cut into lengths of from eight to 10 feet and dragged to the edge of the clearing. The thicker logs are split into rails, and the logs and the rails are lashed together with vines to form the fence. Fence-building is heavy work, even without taking into consideration frequent trips to higher altitudes for the gathering of the strong vines needed for lashing. I estimate that the construction effort alone involved an input of something over 46 kilocalories per running foot of fence. Assuming a need for 370 feet of fence per acre of garden and making allowance for the energy expended in gathering vines, I have calculated that the total input for fencing is 17,082 kilocalories per acre. It is little wonder that the Tsembaga tend to cluster their gardens; clustering reduces the length of fence required per unit of area.

After fences are built, and between one month and four months after clearing begins (depending on how steadily the gardener works and on the weather), the felled litter on the site is burned. This is a step of considerable importance in the swiddening regime. Burning not only disposes of the litter but also liberates the mineral nutrients in the cut vegetation and makes them available to the future crop. Since the layer of fertile soil under tropical forests is remarkably thin (seldom more than two inches in Tsembaga territory) and is easily depleted, the nutrients freed from the fallen trees by burning are beneficial, if not crucial, to the growth of garden plants.

Not much energy is expended in the burning process, although one burning is never enough to finish the job. Moreover, since considerable time usually elapses between the clearing of the underbrush and the burning, it is necessary to weed the garden area before the burning. I estimate that two burnings and a weeding involve an energy input of 9,484 kilocalories per acre.

While the women gather the litter into piles for the second burning, the men put aside some of the lighter unburned logs. Some are laid across the grade of the slope to retain the soil. The rest are used to mark individual garden plots. This task is also light work; I estimate

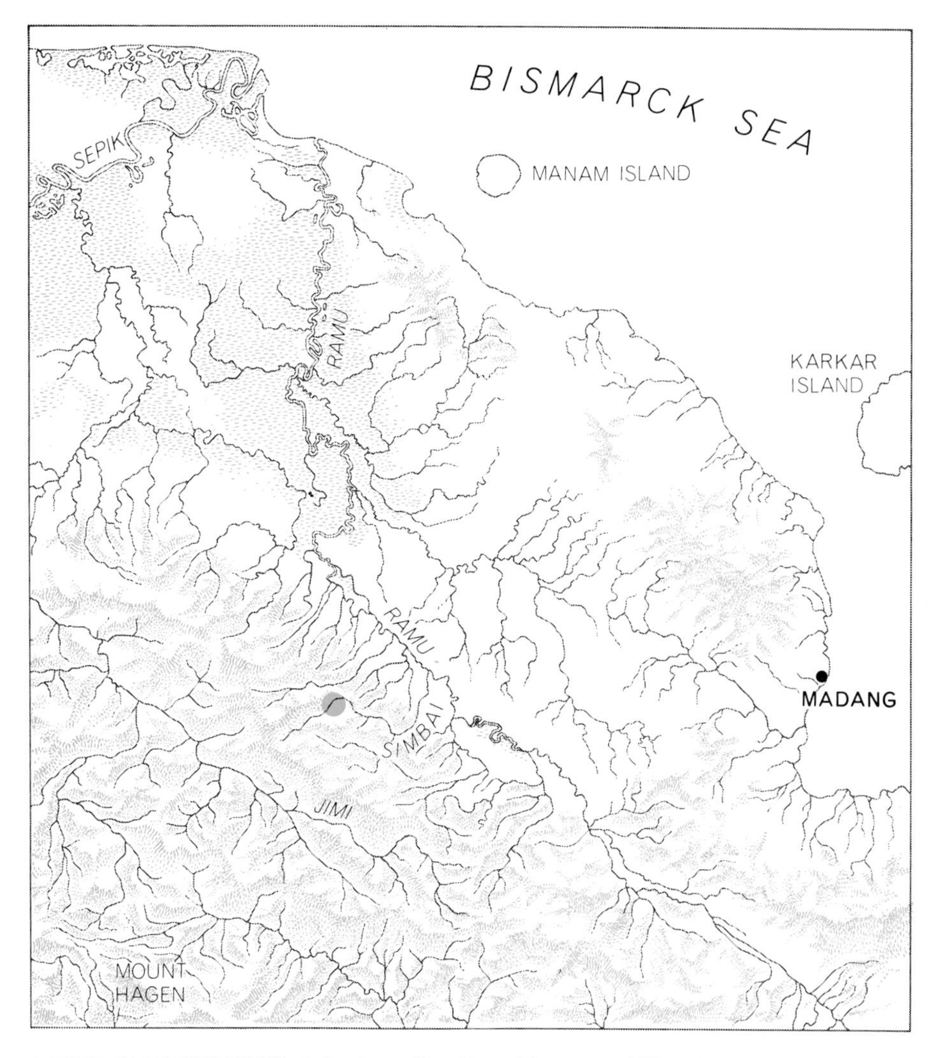

CENTRAL HIGHLANDS of the Australian Trust Territory of New Guinea are drained by the Sepik and Ramu rivers; the Simbai River is a tributary of the Ramu. The colored dot marks the territory of the Tsembaga, the group whose farming practices the author analyzed.

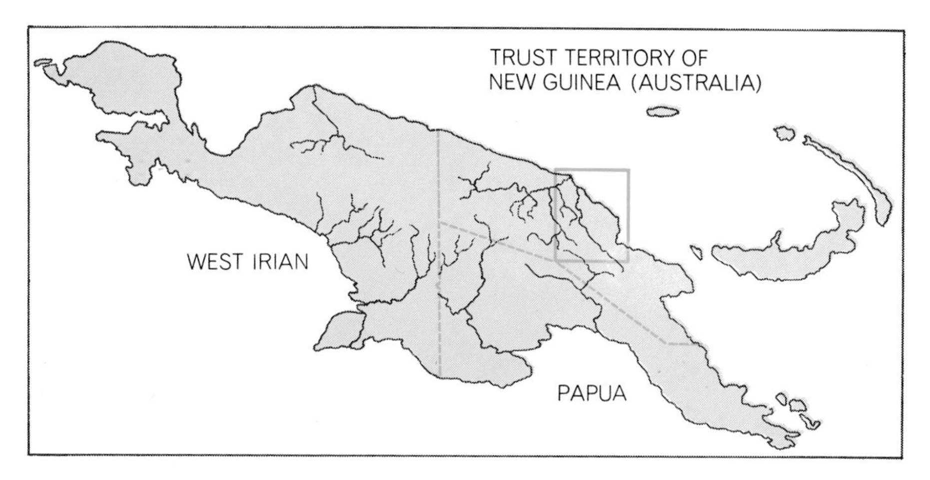

TRUST TERRITORY of New Guinea is one of the island's three political divisions. West Irian (*left*) is a part of the Republic of Indonesia and Papua is an Australian territory.

LITTER IS BURNED in the clearing where a Tsembaga family is preparing to plant a garden. Tree trunks that are not consumed may be used as soil-retainers or plot-markers. Burning releases nutrients in the cut vegetation that are utilized by the garden plants.

that the total energy input is 7,238 kilocalories per acre.

Burning completed and plot-markers and soil-retainers laid, gardens are ready to be planted. For planting stock the Tsembaga depend primarily on cuttings, although they raise a few crops from seed. The gathering and planting of the cuttings, which are set into holes punched in the untilled soil with a heavy stick, is relatively demanding work. I estimate that the men and women who do the planting expend .38 kilocalorie per square foot, or 16,553 kilocalories per acre.

It is appropriate here to list the plants the Tsembaga grow and also to describe the appearance of a growing garden. The Tsembaga can name at least 264 varieties of edible plants, representing some 36 species. The staples are taro and sweet potato. Other starchy vegetables such as yams, cassavas and bananas are of lesser importance. Sweet potatoes and cassavas are used as pig feed as well as for human consumption. Beans, peas, maize and sugarcane are also grown, along with a number of leafy greens, including hibiscus. Hibiscus leaves are in fact the most important plant source of protein in the Tsembaga diet. An asparagus-like plant, *Setaria palmaefolia*, and a relative of sugarcane

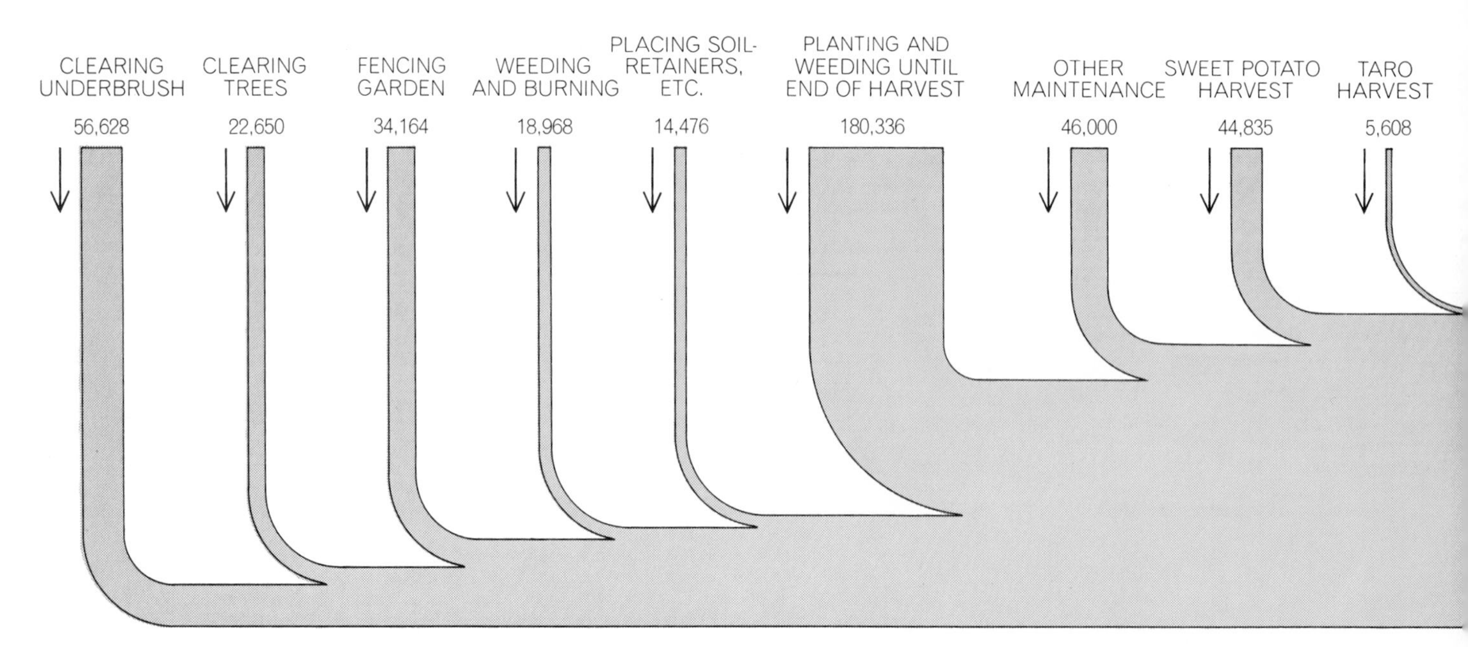

TWELVE MAJOR INPUTS of energy are required in gardening. The flow diagram shows the inputs in terms of the kilocalories per acre required to prepare and harvest a pair of gardens (*see illustration on page 74*). Weeding, a continual process after the garden

NEW CLEARING in second-growth forest contains many stumps of trees that have been cut high for use as props for growing plants. Some, although stripped of their leaves, will survive; along with invading tree seedlings they will slowly reforest the garden site.

known as *pitpit* are valued for their edible flowering parts. So is the *Marita* pandanus, one of the screw pines; its fruit is the source of a thick, fat-rich red fluid that the Tsembaga use as a sauce on greens. Minor garden crops include cucumber, pumpkin, watercress and breadfruit.

Most of the principal crops are to be found in most of the gardens and each is often represented by several varieties of the same species. As Clifford Geertz of the University of Chicago has remarked, there is a structural similarity between a swidden garden and a tropical rain forest. In the garden, as in the forest, species are not segregated by rows or sections but are intricately intermingled, so that as they mature the garden becomes stratified and the plants make maximum use of surface area and of variations in vertical dimensions. For example, taro and sweet potato tubers mature just below the surface; the cassava root lies deeper and yams are the deepest of all. A mat of sweet potato leaves covers the soil at ground level. The taro leaves project above this mat; the hibiscus, sugarcane and *pitpit* stand higher still, and the fronds of the banana spread out above the rest. This intermingling does more than make the best use of a fixed volume. It also dis-

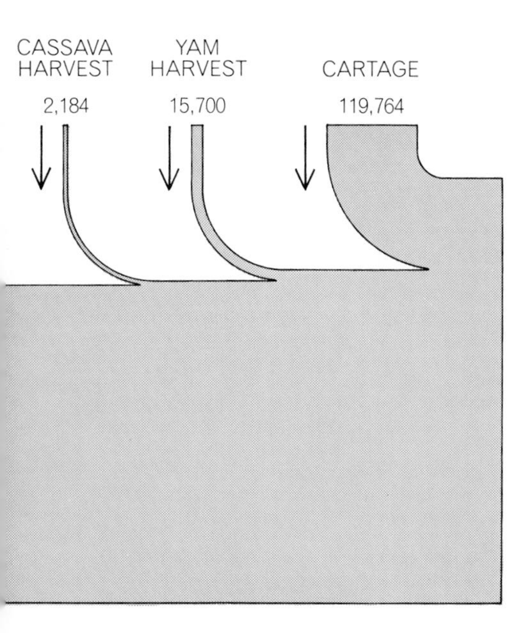

is planted, demands the most energy. Bringing in the garden harvest (*right*) ranks next.

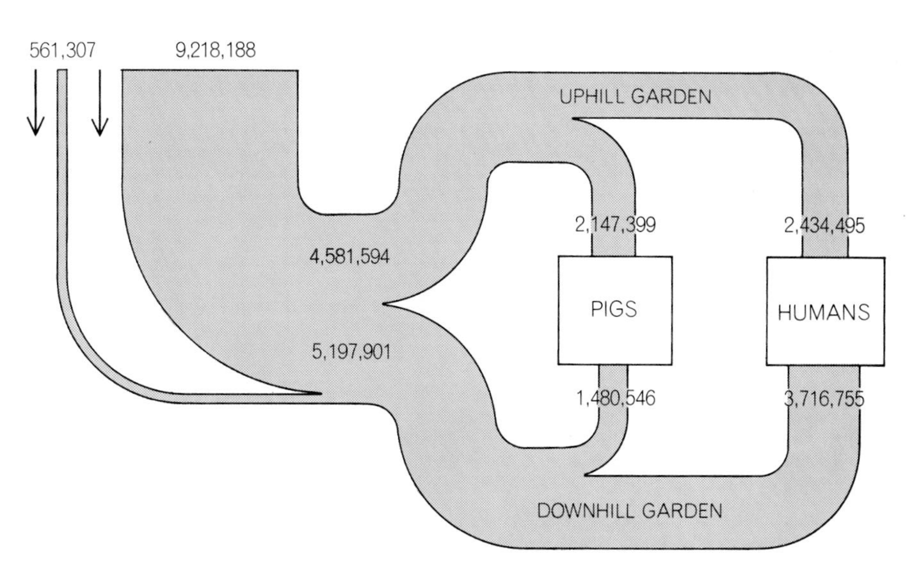

BIOMASS OF CROP YIELD, also measured in kilocalories, gives more than a 16-to-one return on the human energy investment. The Tsembaga use much of the harvest as pig feed.

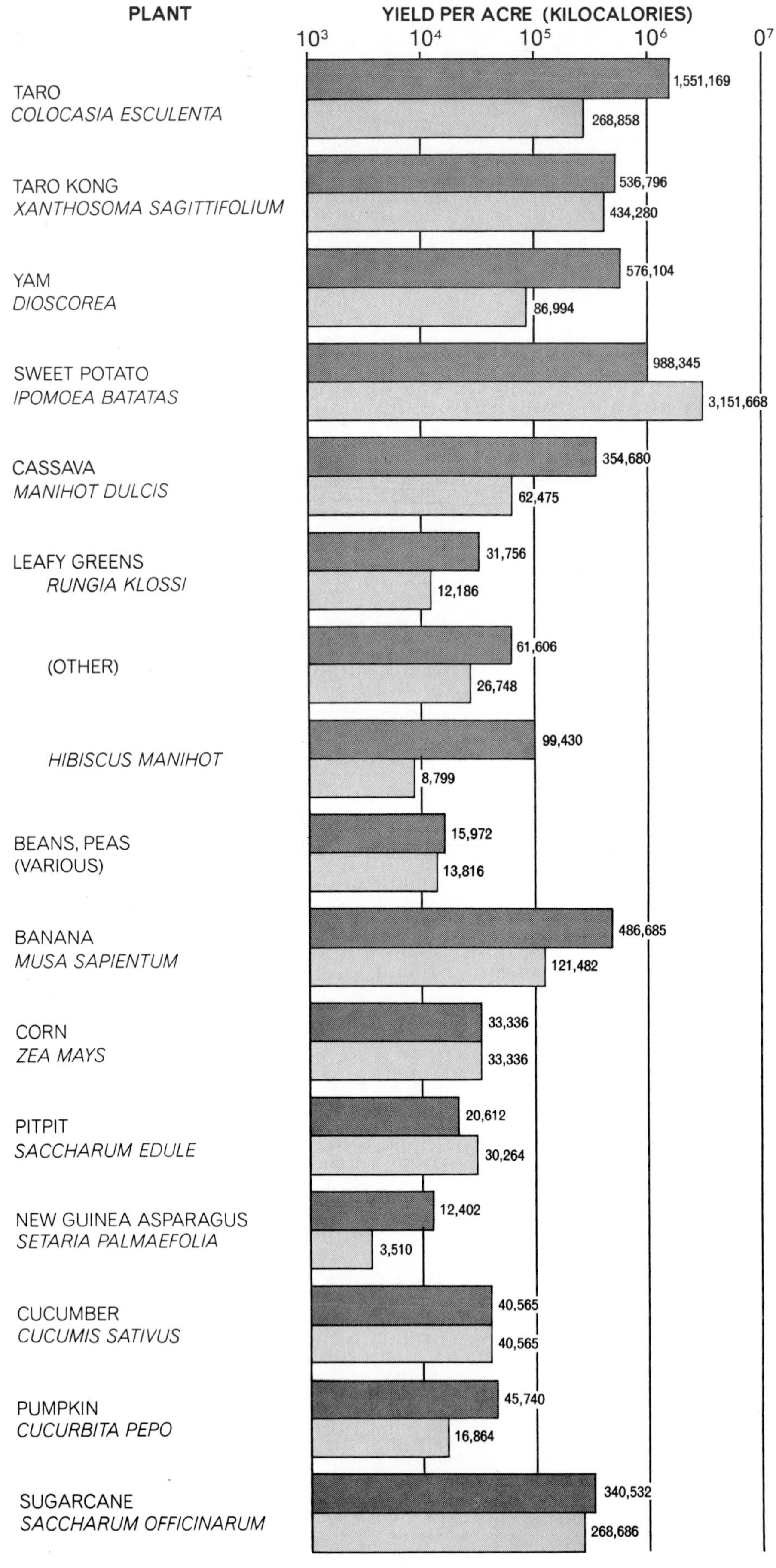

TWO KINDS OF GARDEN are planted by Tsembaga families, whose pigs are too numerous for one garden to feed. Usually one garden is downhill from the other. The downhill garden (*gray*) is planted with more of the family staples: taro and yam. The uphill garden (*color*) is planted with more sweet potato, which is a staple food for pigs. The graph shows the 16 principal Tsembaga garden crops and the comparative yields from two gardens.

courages plant-specific insect pests, it allows advantage to be taken of slight variations in garden habitats, it is protective of the thin tropical soil and it achieves a high photosynthetic efficiency.

If a Tsembaga man and his wife have only one or two pigs, they usually plant one major garden in the middle altitudes, generally between 4,000 and 4,500 feet. But pigs are given a ration from the gardens, and when their numbers increase, their owners are likely to plant two gardens, one above 4,500 feet and the other between 3,000 and 4,000 feet. These gardens differ in the proportions of the major crops with which they are planted. The lower-altitude gardens include more taros and yams and the higher-altitude gardens more sweet potatoes, this last crop forming the largest part of the pigs' ration.

After planting is completed the most laborious of the gardening tasks, weeding, still lies ahead. The Tsembaga recognize and name several successive weedings, but after the first weeding is done (five to seven weeks after planting) the chore becomes virtually continuous. I estimate the total energy input for this task to be about 2.07 kilocalories per square foot, or 90,168 kilocalories per acre. Other miscellaneous tasks performed at the same time (for example tying sugarcane to stumps or other supports) require an additional 14,500 kilocalories per acre. It is noteworthy that weeding aims at uprooting any herbaceous competitors invading the garden but that tree seedlings are spared and even protected. Indeed, a Tsembaga gardener is almost as irritated when a visitor damages a tree seedling as when he heedlessly tramples on a taro plant.

The Tsembaga recognize the importance of the regenerating trees; they call them collectively *duk mi,* or "mother of gardens." Allowing tree seedlings to remain and grow avoids a definite grassy stage in the succession following the abandonment of the garden and for one thing ensures a more rapid redevelopment of the forest canopy. For another, during the cropping period itself the young trees provide a web of roots that penetrate deeper into the ground than the roots of any of the crops and are able to recover nutrients that might otherwise be lost through leaching. Above the ground the developing leaves and branches not only provide additional protection for the thin forest soil against tropical downpours but also immobilize nutrients recovered from the soil for release to future gardens. They also make it increasingly difficult for gardeners to harvest and weed. As a result the people are induced to abandon their gardens before they have seriously depleted the soil, even before the crops are completely harvested. The developing trees, whose growth they themselves have encouraged, make harvesting more laborious at the same time that it is becoming less rewarding. It is interesting to note here that both Ramón Margalef of the University of Barcelona and Howard T. Odum of the University of Florida have argued that in complex ecosystems successful species are not those that merely capture energy more efficiently than their competitors but those that sustain the species supporting them. It is clear that the Tsembaga support not only the garden species that provide them with food but also the species on which they ultimately depend: the species of the forest, which sustain their gardens.

Technological limitations and the nature of crops prevent the Tsembaga from storing most of the food they grow. As a result there is no special harvest period. From the time the crops begin to mature a little harvesting is done every day or two. The strategy is to do as little damage as possible to the plant that yields the crop. For example, hibiscus shrubs are not stripped of all their leaves at one time. Instead every few days a few leaves are plucked from each of several shrubs. This method increases the total yield by allowing the plants to recover in the interval between successive harvestings.

Harvesting continues almost daily from the time crops mature until those in a garden planted in the succeeding year are ready. With the advent of a new garden, harvesting becomes less frequent in the old garden, finally ceasing altogether somewhere between 14 and 28 months after planting. Gardens at lower altitudes are usually shorter-lived than those higher on the mountainside because secondary forest regenerates more quickly on them.

Energy expenditure for harvesting differs slightly between the two types of garden. I estimate that 40,966 kilocalories per acre are expended in harvesting taro-yam gardens and 44,168 kilocalories per acre in sweet potato gardens. Food is consumed at home, homes are at some distance from the gardens and produce therefore must be carried from gardens to houses. An estimated 48,360 kilocalories are expended in transporting produce from sweet potato gardens, which are at approximately the same altitude as the houses. I estimate that 71,404 kilocalories per acre are needed to bring home produce from the taro-yam gardens, which are usually 1,000 to 2,000 feet lower on the mountainside.

FENCE-BUILDING is one of the more energy-consuming tasks in preparing a garden. Fences must be pig-proof to keep out both domestic swine and feral swine from the forest.

Combining all the inputs and comparing input with yield, I found that the Tsembaga received a reasonable short-term return on their investment. The ratio of yield to input was about 16.5 to one for the taro-yam gardens and about 15.9 to one for the sweet potato gardens. Moreover, in 1963, when these observations were made, the distances between the gardens and the residences of the Tsembaga were greater than usual. It was a festival year and the dwellings, instead of being dispersed among the gardens as is customary, were all clustered around a dance ground. If the normal residential pattern had been in effect, garden-to-house distances might have been reduced by as much as 80 percent, and the yield ratios would have risen respectively to 20.1 to one and 18.4 to one.

I have indicated that swine husbandry is intimately related to gardening among the Tsembaga because they devote a substantial proportion of their principal crops to feeding their pigs. Each adult pig receives a daily ration that equals an adult man's ration in weight, although it differs from the human ration in composition. Around the world most pigs are raised to be eaten. This is the ultimate fate that befalls a Tsembaga pig, but such consumption involves social and political relations and religious beliefs and practices, not simply a desire for meat.

The Tsembaga and other groups that speak the Maring language are prevented by strong religious prohibitions from initiating warfare until they have completed a year-long festival that culminates in large-scale sacrifices of pigs to their ancestors, rewarding them for their presumed support in the last round of warfare. The festival itself is the climax of a prolonged ritual cycle that begins years earlier with the sacrifices terminating warfare, sacrifices in which all adult and many juvenile pigs are killed. The Tsembaga held such a festival during my visit in 1962–1963.

When the festival began, the Tsembaga pigs numbered 169 animals with an average per capita weight of from 120 to 150 pounds. This sizable herd was eating 54 percent of all the sweet potatoes and 82 percent of all the cassavas growing in the Tsembaga gardens: some 36 percent of all the tubers of any kind that the Tsembaga grew. The operation called for the commitment of about a third of all garden land to the production of pig feed (not even taking into account household garbage, which the pigs also consumed). At the end of the festival the Tsembaga herd had been reduced to some 60 juvenile pigs, each weighing an average of from 60 to 70 pounds. Thus in terms of live weight the slaughter had decreased the herd sixfold. In anticipation of this decrease the Tsembaga had earlier in 1963 set the area of new land being put into gardens at what amounted to 36.1 percent below the level of the previous year.

For the Tsembaga swine husbandry is obviously an expensive business. The input in human energy, both in growing pig food and in managing the animals, I estimate to be about 45,000 kilocalories per pig per year. Because 10 years or so are needed to bring the herd up from a minimum following the sacrifices terminating warfare to a size large enough for a festival, the ratio of energy yield to energy input in Tsembaga pig husbandry is certainly no better than two to one and is probably worse than one to one. It is evident that keeping pigs cannot be justified or even interpreted in terms of energetics alone. Other possible benefits must also be considered.

First, Maring pig husbandry is part of the means for regulating relations between autonomous local groups such as the Tsembaga. It has already been indicated that the frequency of warfare is regulated by a ritual cycle, but the timing of the cycle is itself a function of the speed of growth in the pig population. Since it usually takes a decade or more to accumulate enough pigs to hold the festival terminating a cycle, the initiation of warfare is held to once per decade or less.

Second, the animals form a link in the detritus food chain by consuming both garbage and the unassimilated vegetable content of human feces, materials that would otherwise be wasted. Moreover, the pigs are regularly penned in abandoned gardens, where they root up unharvested tubers and where, by eliminating herbaceous growth that competes with tree seedlings, they may hasten the return of the forest.

Third, it must be remembered that a diet must provide more than mere energy to a consumer, and it is as converters of carbohydrates of vegetable origin into high-quality protein that pigs are most important in the Tsembaga diet. The importance of their protein contribution is magnified by the circumstances surrounding their consumption.

Except for the once-a-decade festival and for certain rites associated with warfare that are equally infrequent, it is rare for the Tsembaga to kill and eat a pig. The ritual occasions that do call for this unusual behavior are associated with sickness, injury or death. I have argued elsewhere that the sick, the injured, the dying and their kin and associates in Tsembaga society all suffer physiological stress, with an associated net loss of nitrogen (meaning protein) from their body tissues. Now, since the regular protein intake of the Tsembaga is marginal, their antibody production is likely to be low and their rate of recovery from injury or illness slow. In such circumstances nitrogen loss can be a serious matter. The condition of a nitrogen-depleted organism, however, improves rather quickly with the intake of high-quality protein. The nutritional significance of the Tsembaga pigs thus outweighs their high cost in energy input. They provide a source of high-quality protein when it is needed most.

LAYERS OF GROWTH characterize the Tsembaga garden. Root crops develop at different depths and their leaves reach different heights. Bananas (*foreground*) are the tallest plants.

In summary, swiddening and swine husbandry provide the Tsembaga with on the one hand an adequate daily energy ration and on the other an emergency source of protein. The average adult Tsembaga male is four feet 10½ inches tall and weighs 103 pounds; the average adult female is four feet 6½ inches tall and weighs 85 pounds. The garden produce provides the men with some 2,600 kilocalories per day and the women with some 2,200 kilocalories. At the same time the Tsembaga's treatment of the ecosystem in which they live is sufficiently gentle to ensure the continuing regeneration of secondary forest and a continuing supply of fertile garden sites.

We may reflect here on the general strategy of swiddening. It is to establish temporary associations of plants directly useful to man on sites from which forest is removed and to encourage the return of forests to those sites after the useful plants have been harvested. The return of the forest makes it possible, or at least much easier, to establish again an association of cultivated plants sometime in the future. Moreover, the gardens are composed of many varieties of many plant species. If one or another kind of plant succumbs to pests or blight, other plants are available as substitutes. This multiplicity of plants enhances the stability of the Tsembaga subsistence base in exactly the same way that the complexity of the tropical rain forest ensures its stability.

How does such an agricultural pattern compare with farming methods that remove the natural flora over extensive regions and put in their place varieties of one or a few plant species? Before we take up this question let us consider ecosystems in general and the interactions among the participants in ecosystems in particular. All ecosystems are basically similar. Solar energy is photosynthetically fixed by plants—the primary producers—and then passes along food chains that are chiefly made up of animal populations with various feeding habits: herbivores, one or more levels of carnivores and eventually decomposers (which reduce organic wastes into inorganic constituents that can again be taken up by the plants).

Examined in terms of the flow of matter the ecosystem pattern is roughly circular; the same substances are cycled again and again. The flow of energy in ecosystems is linear. It is not recycled but is eventually lost to the local system.

The energy supply is steadily degraded into heat with each successive step along the food chain. For example, of the organic material synthesized by a plant from 80 to 90 percent is available under ordinary circumstances to the herbivore that consumes the plant. The herbivore expends energy, however, not only in feeding and in other activities but also in metabolic processes. When the amount of energy captured by one organism in a food chain is compared with the amount of energy the organism yields to the organism that preys on it, the figure is usually of the order of 10 to one, which means an energy loss of 90 percent.

Just as there are general similarities among ecosystems, so there are differences in the numbers and the kinds of species that participate in each system and in the relations among them. These differences are the result of evolutionary processes. Associations of species adapt to their habitat, and with the passage of time they pass through more or less distinct stages known as successions. As the work of Margalef and of Eugene P. Odum of the University of Georgia has demonstrated, these successions also show certain general similarities.

GARDEN MAINTENANCE includes light tasks such as tying up sugarcane. Here the stalks of cane *(left)* have been lashed to a tree stump by a gardener with a length of vine.

One similarity is that in any ecosystem the total biomass increases as time passes. There is a corresponding but not linearly related increase in primary productivity—that is, photosynthesis—whether it is expressed as the total amount of energy fixed or as the total quantity of organic material synthesized. What allows the increase in primary productivity is a parallel increase in the number of plant species that have become specialized and are thus able to conduct photosynthesis under a greater variety of circumstances in a wider range of microhabitats. The increase in plant biomass and plant productivity may allow some increase in animal biomass too. It certainly encourages an increase in the diversity of animal species, since the increase in the diversity of plants favors specialization among herbivores.

Not only are there more species in the maturer stages of a succession but also the species are likely to be of a different kind. Species typical of immature succession stages—"pioneer" species—are characteristically able to disperse themselves over considerable distances, to reproduce prodigally, to compete strongly for dominance and to survive under unstable and even violently fluctuating conditions. They also tend to be short-lived, that is, their populations are quickly replaced. Therefore in immature ecosystems the ratio of productivity to biomass is high.

In contrast, the species characteristic of more mature stages of succession often cannot disperse themselves easily over long distances, produce few offspring and are relatively long-lived. Their increasing specialization militates against overt competition among them; instead the relations among species are often characterized by an increased mutual reliance. In effect, as the ecosystem becomes maturer it becomes more complex for the number of species, and their interdependence increases. As its complexity increases so does its stability, since there are increasing numbers of alternative paths through which energy and materials can flow. Therefore if one or another species is decimated, the entire system is not necessarily endangered.

The maturing ecosystem also becomes more efficient. In the immature stage productivity per unit of biomass is high and total biomass is likely to be low. As stage follows stage both productivity and biomass increase, but there is a decrease in the ratio of productivity to biomass. For example, it takes less energy to support a pound of biomass in a mature tropical rain forest than it does in the grassy or scrubby forest stages that precede maturity.

It is not really surprising that increased stability and increased efficiency should be characteristic of mature ecosystems. Some of the implications of this trend, however, are not altogether obvious. I have alluded to the arguments of Margalef and Howard Odum suggesting that in mature ecosystems selection is not merely a matter of competition between individuals or species occupying the same level in the food chain. Rather, it favors those species that contribute positively to the stability and efficiency of the entire ecosystem. Organisms with positions well up along the food chain can "reward" the organisms they depend on by returning to these lower organisms materials they require or by performing services that are beneficial to them. An exploitive population—one that consumes at a rate greater than the productivity of the species it depends on or does not reward that species—will eventually perish.

Margalef and others have pointed out that human intervention tends to reduce the maturity of ecosystems. Both in terms of the small number of species present and of the lack of ecological complexity a farm or a plantation more closely resembles an immature stage of succession than it does a mature stage. Furthermore, in ecosystems dominated by man the chosen species are usually quick to ripen—that is, they are short-lived—and the productivity per unit of biomass is likely to be high. Yet there is a crucial difference between natural pioneer associations and those dominated by man.

Pioneers in an immature ecosystem are characterized by their ability to survive under unstable conditions. Man's favored cultigens, however, are seldom if ever notable for hardiness and self-sufficiency. Some are ill-adapted to their surroundings, some cannot even propagate themselves without assistance and some are able to survive only if they are constantly protected from the competition of the natural pioneers that promptly invade the simplified ecosystems man has constructed. Indeed, in man's quest for higher plant yields he has devised some of the most delicate and unstable ecosystems ever to have appeared on the face of the earth. The ultimate in human-dominated associations are fields planted in one high-yielding variety of a single species. It is apparent that in the ecosystems dominated by man the trend of what can be called successive anthropocentric stages is exactly the reverse of the trend in natural ecosystems. The anthropocentric trend is in the direction of simplicity rather than complexity, of fragility rather than stability.

We return to our question of how the pattern of Tsembaga agriculture compares with the pattern of agriculture based on one crop. The question really concerns the interaction of man's expanding social, political and economic organization on the one hand and local natural ecosystems on the other. The trend of anthropocentric succession can best be understood as one aspect of the evolution of human society.

We can place the Tsembaga and other "primitive" horticulturists early in such successions. The Tsembaga are politically and economically autonomous. They neither import nor export foodstuffs, and it is necessary for them to maintain as wide a range of crops in their gardens as they wish to enjoy. Economic self-sufficiency obviously encourages a generalized horticulture, but economic self-sufficiency and production for use protect ecological integrity in more

FERAL PIG (*foreground*) has been roasted and is being dismembered. Feral animals are eaten whenever they are killed, but domestic swine are killed only on ceremonial occasions.

important ways. For one thing, the management of cultivation is only slightly concerned, if at all, with events in the outside world. On the contrary, it attends almost exclusively to the needs of the cultivators and the species sustaining them. Moreover, in the absence of exotic energy sources the ability of humans to abuse the species on which they depend is limited by those species, because it is only from them that energy for work can be derived. In such systems, however, abuses seldom need to be repaid by declining yields because they are quickly signaled to those responsible by subtler signs of environmental degradation. Information feedback from the environment is sensitive and rapid in small autonomous ecological systems, and such systems are likely to be rapidly self-correcting.

The fact remains that autonomous local ecological systems such as the Tsembaga system have virtually disappeared. All but a few have been absorbed into an increasingly differentiated and complex social and economic organization of worldwide scope. The increasing size and complexity of human organization is related to man's increasing ability to harness energy. The relationship is not simple; rather it is one of mutual causation. As an example, increases in the available energy allow increases in the size and differentiation of human societies. Increased numbers and increasingly complex organizations require still more energy to sustain them and at the same time facilitate the development of new techniques for capturing more energy, and so on. The system is characterized by positive feedback.

Leslie A. White of the University of Michigan suggested that cultural evolution can be measured in terms of the increasing amount of energy harnessed per capita per annum. This is a yardstick that seems generally to agree with the historical experience of mankind. The development of energy sources that are independent of immediate biological processes has been the factor of greatest importance. Industrialized societies harness many times more energy per capita per annum than nonindustrial ones, and the energy-rich have enjoyed a great advantage in their relations with the energy-poor. In areas where the two have competed the nonindustrial societies have inevitably been displaced, absorbed or destroyed.

Agriculture is not exempt from the increasing specialization that characterizes social evolution generally. Not only individual farms but also regions

COSTUMED FOR DANCING, a group of Tsembaga men stands ready near the dance ground. During this ceremony in 1963 the Tsembaga butchered and distributed 109 pigs.

and even nations have been turned into man-made immature ecosystems such as cotton plantations or cane fields. It is important to note that the transformation would be virtually impossible without sources of energy other than local biological processes. Fossil fuels come into play. When such energy sources are available, the pressures that can be brought to bear on specific ecosystems are no longer limited to the energy that the ecosystem itself can generate, and alterations become feasible that were formerly out of reach. A farmer may even expend more energy in the gasoline consumed by his farm machinery than is returned by the crop he raises. The same nonbiological power sources make it possible to provide the world agricultural community with the large quantities of pesticides, fertilizers and other kinds of assistance that many man-made immature ecosystems require in order to remain productive. Moreover, the entire infrastructure of commercial agriculture—high-speed transportation and communications, large-scale storage facilities and elaborate economic institutions—depends on these same sources of nonbiological energy.

As man forces the ecosystems he dominates to be increasingly simple, however, their already limited autonomy is further diminished. They are subject not only to local environmental stress but also to extraneous economic and political vicissitudes. They come to rely more and more on imported materials; the men who manipulate them become more and more subject to distant events, interests and processes that they may not even grasp and certainly do not control. National and international concerns replace local considerations, and with the regulation of the local ecosystems coming from outside, the system's normal self-corrective capacity is diminished and eventually destroyed.

Margalef has observed that in exchanges among systems differing in complexity of organization the flow of material, information and energy is usually from the less highly organized to the more highly organized. This principle may find expression not only in the relations between predator and prey but also in the relations between "developed" and "underdeveloped" nations. André Gunder Frank has argued that in the course of the development of underdeveloped agrarian societies by industrialized societies the flow of wealth is usually from the former to the latter. Be this as it may, economic development surely accelerates ecological simplification; it inevitably encourages a shift from more diverse subsistence agriculture to the cultivation of a few crops for sale in a world market.

It may not be improper to characterize as ecological imperialism the elaboration of a world organization that is centered in industrial societies and degrades the ecosystems of the agrarian societies it absorbs. Ecological imperialism is in some ways similar to economic imperialism. In both there is a flow of energy and material from the less organized system to the more organized one, and both may simply be different aspects of the same relations. Both may also be masked by the same euphemisms, among which "progress" and "development" are prominent.

The anthropocentric trend I have described may have ethical implications, but the issue is ultimately not a matter of morality or even of *Realpolitik*. It is one of biological viability. The increasing scope of world organization and the increasing industrialization and energy consumption on which it depends have been taken by Western man virtually to define social evolution and progress. It must be remembered that man is an animal, that he survives biologically or not at all, and that his biological survival, like that of all animals, requires the survival of the other species on which he depends. The general ecological perspective outlined here suggests that some aspects of what we have called progress or social evolution may be maladaptive. We may ask if a worldwide human organization can persist and elaborate itself indefinitely at the expense of decreasing the stability of its own ecological foundations. We cannot and would not want to return to a world of autonomous ecosystems such as the Tsembaga's; in such systems all men and women are and must be farmers. We may ask, however, if the chances for human survival might not be enhanced by reversing the modern trend of successions in order to increase the diversity and stability of local, regional and national ecosystems, even, if need be, at the expense of the complexity and interdependence of worldwide economic organization. It seems to me that the trend toward decreasing ecosystemic complexity and stability, rather than threats of pollution, overpopulation or even energy famine, is the ultimate ecological problem immediately confronting man. It also may be the most difficult to solve, since the solution cannot easily be reconciled with the values, goals, interests and political and economic institutions prevailing in industrialized and industrializing nations.

7

The Flow of Energy in an Industrial Society

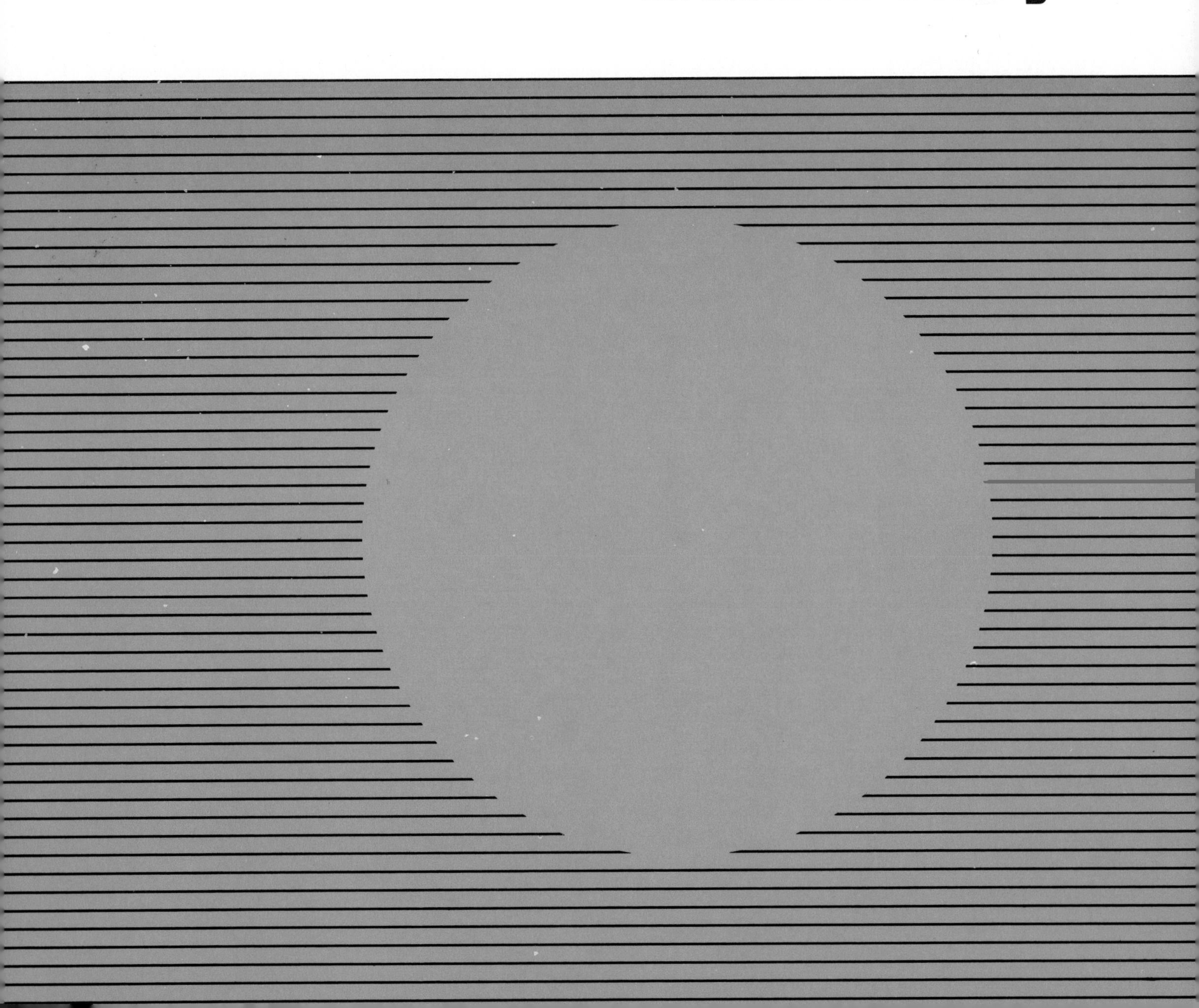

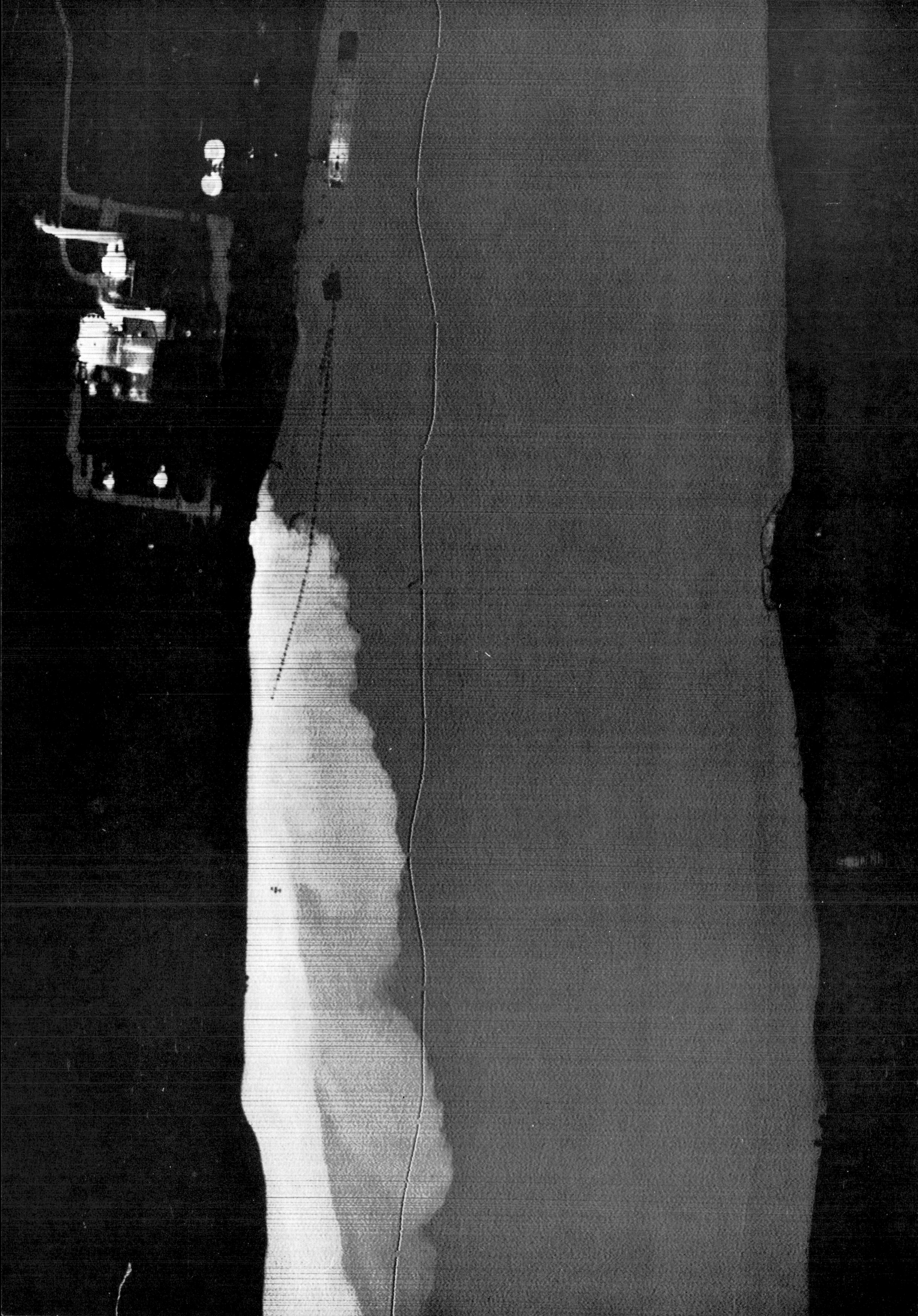

The Flow of Energy in an Industrial Society

EARL COOK

The U.S., with 6 percent of the world's population, uses 35 percent of the world's energy. In the long run the limiting factor in high levels of energy consumption will be the disposal of the waste heat

This article will describe the flow of energy through an industrial society: the U.S. Industrial societies are based on the use of power: the rate at which useful work is done. Power depends on energy, which is the ability to do work. A power-rich society consumes—more accurately, degrades—energy in large amounts. The success of an industrial society, the growth of its economy, the quality of the life of its people and its impact on other societies and on the total environment are determined in large part by the quantities and the kinds of energy resources it exploits and by the efficiency of its systems for converting potential energy into work and heat.

Whether by hunting, by farming or by burning fuel, man introduces himself into the natural energy cycle, converting energy from less desired forms to more desired ones: from grass to beef, from wood to heat, from coal to electricity. What characterizes the industrial societies is their enormous consumption of energy and the fact that this consumption is primarily at the expense of "capital" rather than of "income," that is, at the expense of solar energy stored in coal, oil and natural gas rather than of solar radiation, water, wind and muscle power. The advanced industrial societies, the U.S. in particular, are further characterized by their increasing dependence on electricity, a trend that has direct effects on gross energy consumption and indirect effects on environmental quality.

The familiar exponential curve of increasing energy consumption can be considered in terms of various stages of human development [*see illustration on next page*]. As long as man's energy consumption depended on the food he could eat, the rate of consumption was some 2,000 kilocalories per day; the domestication of fire may have raised it to 4,000 kilocalories. In a primitive agricultural society with some domestic animals the rate rose to perhaps 12,000 kilocalories; more advanced farming societies may have doubled that consumption. At the height of the low-technology industrial revolution, say between 1850 and 1870, per capita daily consumption reached 70,000 kilocalories in England, Germany and the U.S. The succeeding high-technology revolution was brought about by the central electric-power station and the automobile, which enable the average person to apply power in his home and on the road. Beginning shortly before 1900, per capita energy consumption in the U.S. rose at an increasing rate to the 1970 figure: about 230,000 kilocalories per day, or about 65×10^{15} British thermal units (B.t.u.) per year for the country as a whole. Today the industrial regions, with 30 percent of the world's people, consume 80 percent of the world's energy. The U.S., with 6 percent of the people, consumes 35 percent of the energy.

In the early stages of its development in western Europe industrial society based its power technology on income sources of energy, but the explosive growth of the past century and a half has been fed by the fossil fuels, which are not renewable on any time scale meaningful to man. Modern industrial society is totally dependent on high rates of consumption of natural gas, petroleum and coal. These nonrenewable fossil-fuel resources currently provide 96 percent of the gross energy input into the U.S. economy [*see top illustration on page* 85]. Nuclear power, which in 1970 accounted for only .3 percent of the total energy input, is also (with present reactor technology) based on a capital source of energy: uranium 235. The energy of falling water, converted to hydropower, is the only income source of energy that now makes any significant contribution to the U.S. economy, and its proportional role seems to be declining from a peak reached in 1950.

Since 1945 coal's share of the U.S. energy input has declined sharply, while both natural gas and petroleum have increased their share. The shift is reflected in import figures. Net imports of petroleum and petroleum products doubled between 1960 and 1970 and now constitute almost 30 percent of gross consumption. In 1960 there were no imports of natural gas; last year natural-gas imports (by pipeline from Canada and as liquefied gas carried in cryogenic tankers) accounted for almost 4 percent of gross consumption and were increasing.

The reasons for the shift to oil and gas are not hard to find. The conversion of railroads to diesel engines represented a large substitution of petroleum for coal. The rapid growth, beginning during World War II, of the national

HEAT DISCHARGE from a power plant on the Connecticut River at Middletown, Conn., is shown in this infrared scanning radiograph. The power plant is at upper left, its structures outlined by their heat radiation. The luminous cloud running along the left bank of the river is warm water discharged from the cooling system of the plant. The vertical oblong object at top left center is an oil tanker. The luminous spot astern is the infrared glow of its engine room. The dark streak between the tanker and the warm-water region is a breakwater. The irregular line running down the middle of the picture is an artifact of the infrared scanning system. The picture was made by HRB-Singer, Inc., for U.S. Geological Survey.

network of high-pressure gas-transmission lines greatly extended the availability of natural gas. The explosion of the U.S. automobile population, which grew twice as fast as the human population in the decade 1960–1970, and the expansion of the nation's fleet of jet aircraft account for much of the increase in petroleum consumption. In recent years the demand for cleaner air has led to the substitution of natural gas or low-sulfur residual fuel oil for high-sulfur coal in many central power plants.

An examination of energy inputs by sector of the U.S. economy rather than by source reveals that much of the recent increase has been going into household, commercial and transportation applications rather than industrial ones [*see bottom illustration on opposite page*]. What is most striking is the growth of the electricity sector. In 1970 almost 10 percent of the country's useful work was done by electricity. That is not the whole story. When the flow of energy from resources to end uses is charted for 1970 [*see illustration on pages 86 and 87*], it is seen that producing that much electricity accounted for 26 percent of the gross consumption of energy, because of inefficiencies in generation and transmission. If electricity's portion of end-use consumption rises to about 25 percent by the year 2000, as is expected, then its generation will account for between 43 and 53 percent of the country's gross energy consumption. At that point an amount of energy equal to about half of the useful work done in the U.S. will be in the form of waste heat from power stations!

All energy conversions are more or less inefficient, of course, as the flow diagram makes clear. In the case of electricity there are losses at the power plant, in transmission and at the point of application of power; in the case of fuels consumed in end uses the loss comes at the point of use. The 1970 U.S. gross consumption of 64.6×10^{15} B.t.u. of energy (or 16.3×10^{15} kilocalories, or 19×10^{12} kilowatt-hours) ends up as 32.8×10^{15} B.t.u. of useful work and 31.8×10^{15} B.t.u. of waste heat, amounting to an overall efficiency of about 51 percent.

The flow diagram shows the pathways of the energy that drives machines, provides heat for manufacturing processes and heats, cools and lights the country. It does not represent the total energy budget because it includes neither food nor vegetable fiber, both of which bring solar energy into the economy through photosynthesis. Nor does it include environmental space heating by solar radiation, which makes life on the earth possible and would be by far the largest component of a total energy budget for any area and any society.

The minute fraction of the solar flux that is trapped and stored in plants provides each American with some 10,000 kilocalories per day of gross food production and about the same amount in the form of nonfood vegetable fiber. The fiber currently contributes little to the energy supply. The food, however, fu-

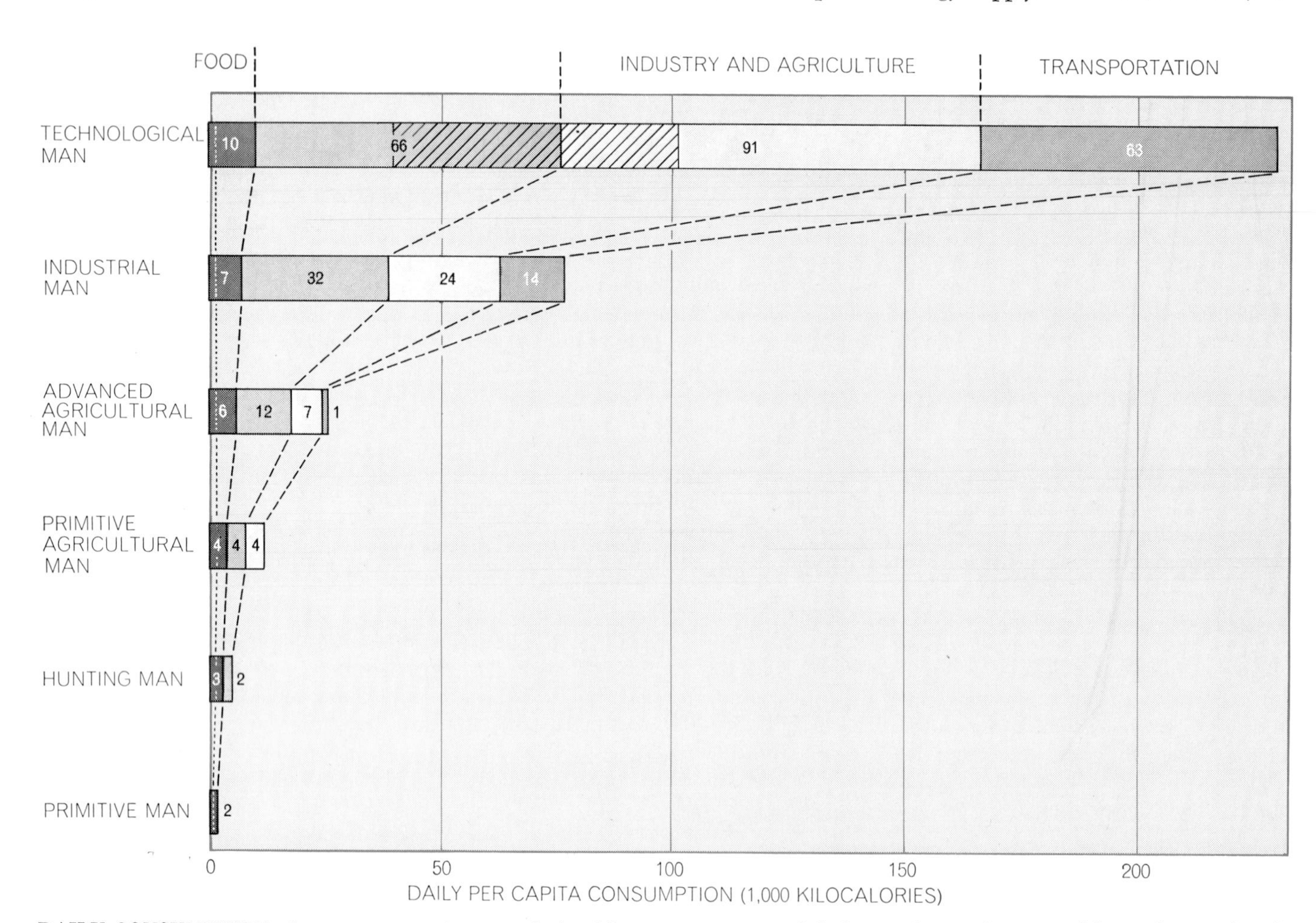

DAILY CONSUMPTION of energy per capita was calculated by the author for six stages in human development (and with an accuracy that decreases with antiquity). Primitive man (East Africa about 1,000,000 years ago) without the use of fire had only the energy of the food he ate. Hunting man (Europe about 100,000 years ago) had more food and also burned wood for heat and cooking. Primitive agricultural man (Fertile Crescent in 5000 B.C.) was growing crops and had gained animal energy. Advanced agricultural man (northwestern Europe in A.D. 1400) had some coal for heating, some water power and wind power and animal transport. Industrial man (in England in 1875) had the steam engine. In 1970 technological man (in the U.S.) consumed 230,000 kilocalories per day, much of it in form of electricity (*hatched area*). Food is divided into plant foods (*far left*) and animal foods (or foods fed to animals).

els man. Gross food-plant consumption might therefore be considered another component of gross energy consumption; it would add about 3×10^{15} B.t.u. to the input side of the energy-flow scheme. Of the 10,000 kilocalories per capita per day of gross production, handling and processing waste 15 percent. Of the remaining 8,500 kilocalories, some 6,300 go to feed animals that produce about 900 kilocalories of meat and 2,200 go into the human diet as plant materials, for a final food supply of about 3,100 kilocalories per person. Thus from field to table the efficiency of the food-energy system is 31 percent, close to the efficiency of a central power station. The similarity is not fortuitous; in both systems there is a large and unavoidable loss in the conversion of energy from a less desired form to a more desired one.

Let us consider recent changes in U.S. energy flow in more detail by seeing how the rates of increase in various sectors compare. Not only has energy consumption for electric-power generation been growing faster than the other sectors but also its growth rate has been increasing: from 7 percent per year in 1961–1965 to 8.6 percent per year in 1965–1969 to 9.25 percent last year [*see top illustration on page 88*]. The energy consumed in industry and commerce and in homes has increased at a fairly steady rate for a decade, but the energy demand of transportation has risen more sharply since 1966. All in all, energy consumption has been increasing lately at a rate of 5 percent per year, or four times faster than the increase in the U.S. population. Meanwhile the growth of the gross national product has tended to fall off, paralleling the rise in energy sectors other than fast-growing transportation and electricity. The result is a change in the ratio of total energy consumption to G.N.P. [*see bottom illustration on page 88*]. The ratio had been in a long general decline since 1920 (with brief reversals) but since 1967 it has risen more steeply each year. In 1970 the U.S. consumed more energy for each dollar of goods and services than at any time since 1951.

Electricity accounts for much of this decrease in economic efficiency, for several reasons. For one thing, we are substituting electricity, with a thermal efficiency of perhaps 32 percent, for many direct fuel uses with efficiencies ranging from 60 to 90 percent. Moreover, the fastest-growing segment of end-use consumption has been electric air conditioning. From 1967 to 1970 consumption for

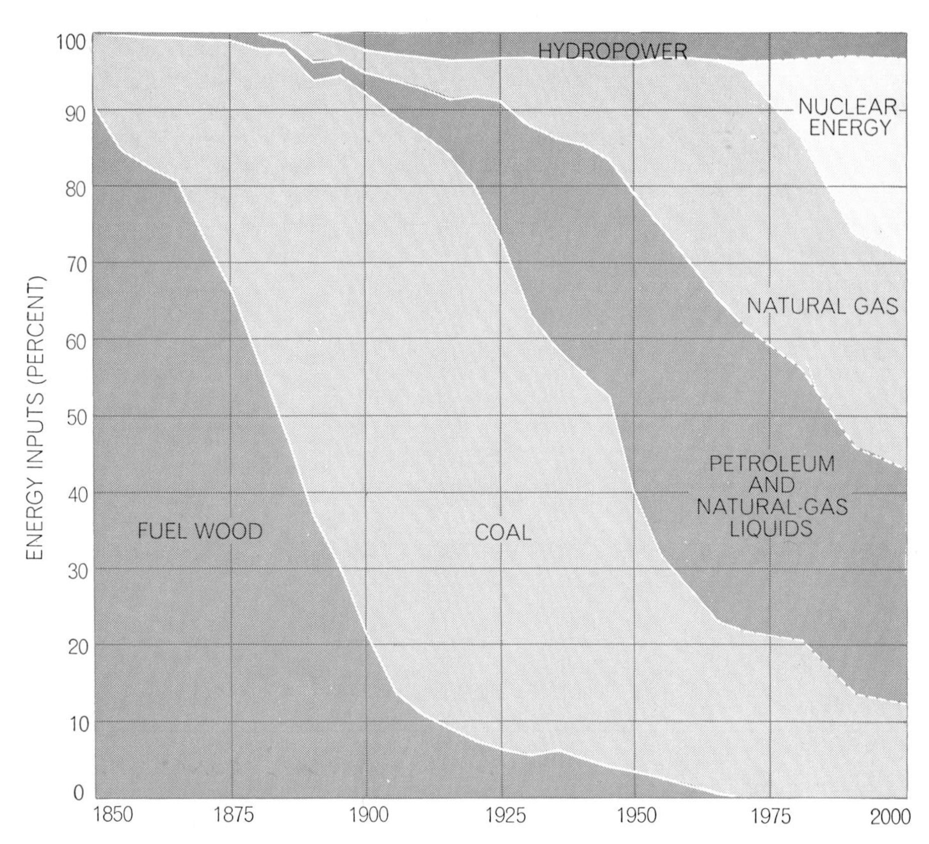

FOSSIL FUELS now account for nearly all the energy input into the U.S. economy. Coal's contribution has decreased since World War II; that of natural gas has increased most in that period. Nuclear energy should contribute a substantial percent within the next 20 years.

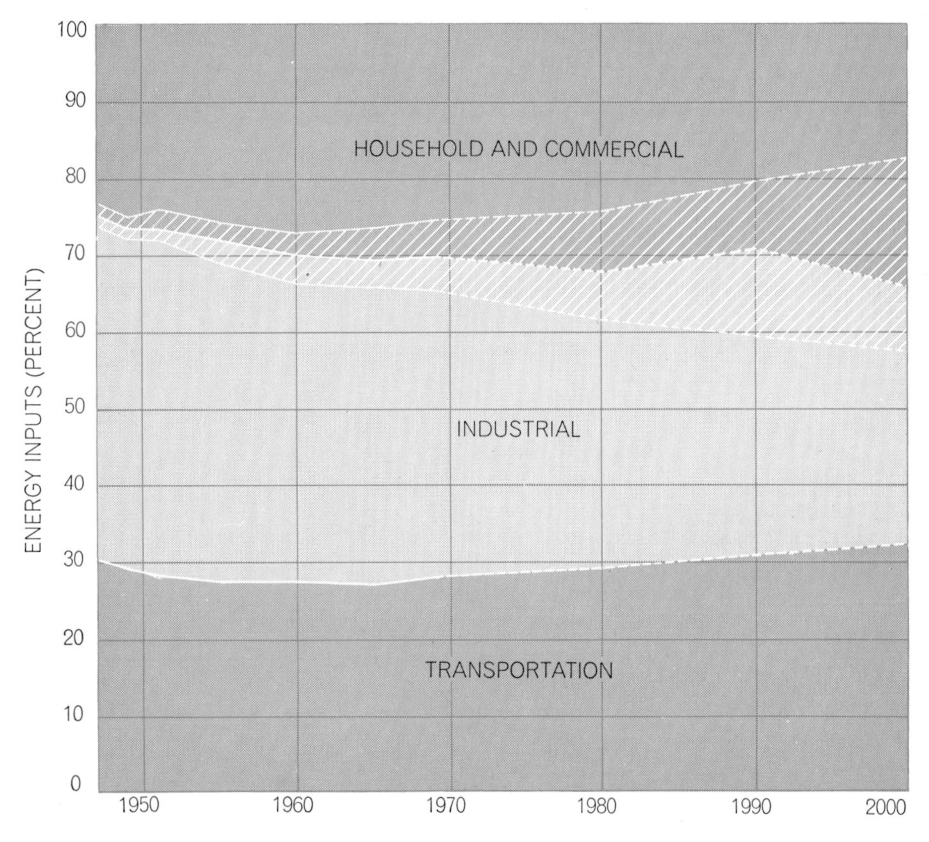

USEFUL WORK is distributed among the various end-use sectors of the U.S. economy as shown. The trend has been for industry's share to decrease, with household and commercial uses (including air conditioning) and transportation growing. Electricity accounts for an ever larger share of the work (*hatched area*). U.S. Bureau of Mines figures in this chart include nonenergy uses of fossil fuels, which constitute about 7 percent of total energy inputs.

air conditioning grew at the remarkable rate of 20 percent per year; it accounted for almost 16 percent of the total increase in electric-power generation from 1969 to 1970, with little or no multiplier effect on the G.N.P.

Let us take a look at this matter of efficiency in still another way: in terms of useful work done as a percentage of gross energy input. The "useful-work equivalent," or overall technical efficiency, is seen to be the product of the conversion efficiency (if there is an intermediate conversion step) and the application efficiency of the machine or device that does the work [*see bottom illustration on page 89*]. Clearly there is a wide range of technical efficiencies in energy systems, depending on the conversion devices. It is often said that electrical resistance heating is 100 percent efficient, and indeed it is in terms, say, of converting electrical energy to thermal energy at the domestic hot-water heater. In terms of the energy content of the natural gas or coal that fired the boiler that made the steam that drove the turbine that turned the generator that produced the electricity that heated the wires that warmed the water, however, it is not so efficient.

The technical efficiency of the total U.S. energy system, from potential energy at points of initial conversion to work at points of application, is about 50 percent. The economic efficiency of

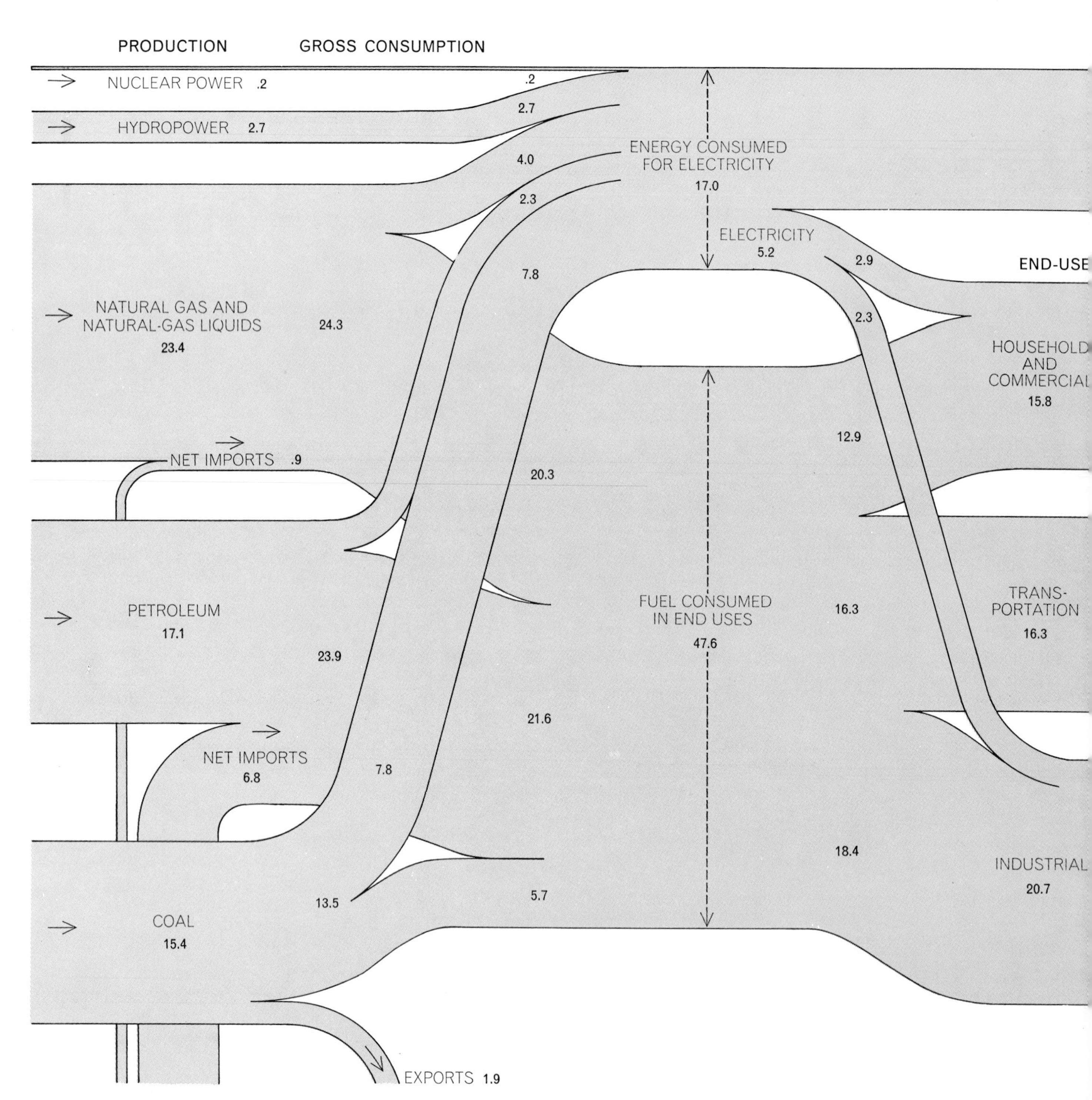

FLOW OF ENERGY through the U.S. system in 1970 is traced from production of energy commodities (*left*) to the ultimate conversion of energy into work for various industrial end products and waste heat (*right*). Total consumption of energy in 1970 was 64.6×10^{15} British thermal units. (Adding nonenergy uses of fossil fuels, primarily for petrochemicals, would raise the total to 68.8×10^{15} B.t.u.) The overall efficiency of the system was about 51 percent. Some of the fossil-fuel energy is consumed directly and

the system is considerably less. That is because work is expended in extracting, refining and transporting fuels, in the construction and operation of conversion facilities, power equipment and electricity-distribution networks, and in handling waste products and protecting the environment.

An industrial society requires not only a large supply of energy but also a high use of energy per capita, and the society's economy and standard of living are shaped by interrelations among resources, population, the efficiency of conversion processes and the particular applications of power. The effect of these interrelations is illustrated by a comparison of per capita energy consumption and per capita output for a number of countries [*see illustration on page 90*]. As one might expect, there is a strong general correlation between the two measures, but it is far from being a one-to-one correlation. Some countries (the U.S.S.R. and the Republic of South Africa, for example) have a high energy consumption with respect to G.N.P.; other countries (such as Sweden and New Zealand) have a high output with relatively less energy consumption. Such differences reflect contrasting combinations of energy-intensive heavy industry and light consumer-oriented and service industries (characteristic of different stages of economic development) as well as differences in the efficiency of energy use. For example, countries that still rely on coal for a large part of their energy requirement have higher energy inputs per unit of production than those that use mainly petroleum and natural gas.

A look at trends from the U.S. past is also instructive. Between 1800 and 1880 total energy consumption in the U.S. lagged behind the population increase, which means that per capita energy consumption actually declined somewhat. On the other hand, the American standard of living increased during this period because the energy supply in 1880 (largely in the form of coal) was being used much more efficiently than the energy supply in 1800 (largely in the form of wood). From 1900 to 1920 there was a tremendous surge in the use of energy by Americans but not a parallel increase in the standard of living. The ratio of energy consumption to G.N.P. increased 50 percent during these two decades because electric power, inherently less efficient, began being substituted for the direct use of fuels; because the automobile, at best 25 percent efficient, proliferated (from 8,000 in 1900 to 8,132,000 in 1920), and because mining and manufacturing, which are energy-intensive, grew at very high rates during this period.

Then there began a long period during which increases in the efficiency of energy conversion and utilization fulfilled about two-thirds of the total increase in demand, so that the ratio of energy consumption to G.N.P. fell to about 60 percent of its 1920 peak although per capita energy consumption continued to increase. During this period (1920–1965) the efficiency of electric-power generation and transmission almost trebled, mining and manufacturing grew at much lower rates and the services sector of the economy, which is not energy-intensive, increased in importance.

"Power corrupts" was written of man's control over other men but it applies also to his control of energy re-

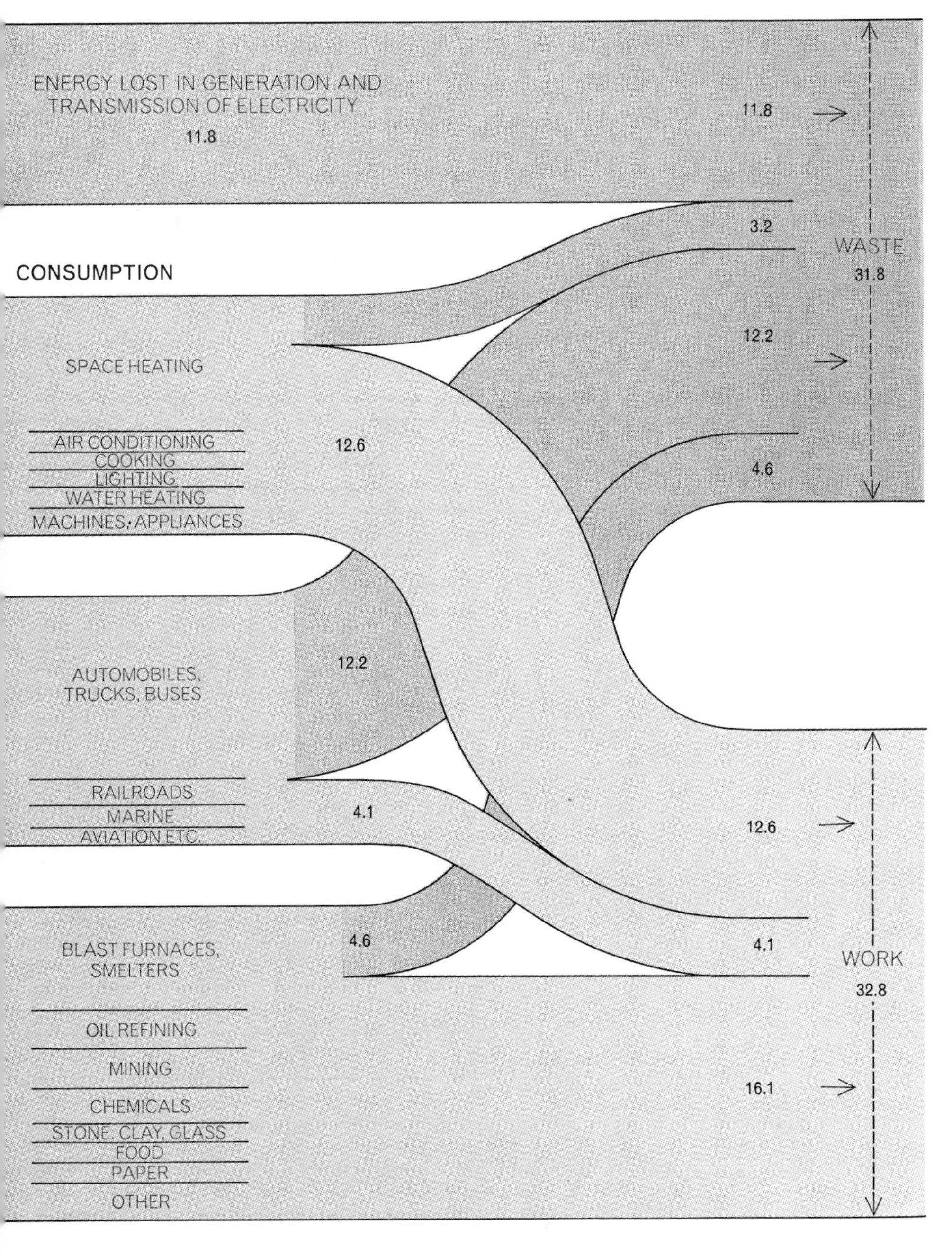

some is converted to generate electricity. The efficiency of electrical generation and transmission is taken to be about 31 percent, based on the ratio of utility electricity purchased in 1970 to the gross energy input for generation in that year. Efficiency of direct fuel use in transportation is taken as 25 percent, of fuel use in other applications as 75 percent.

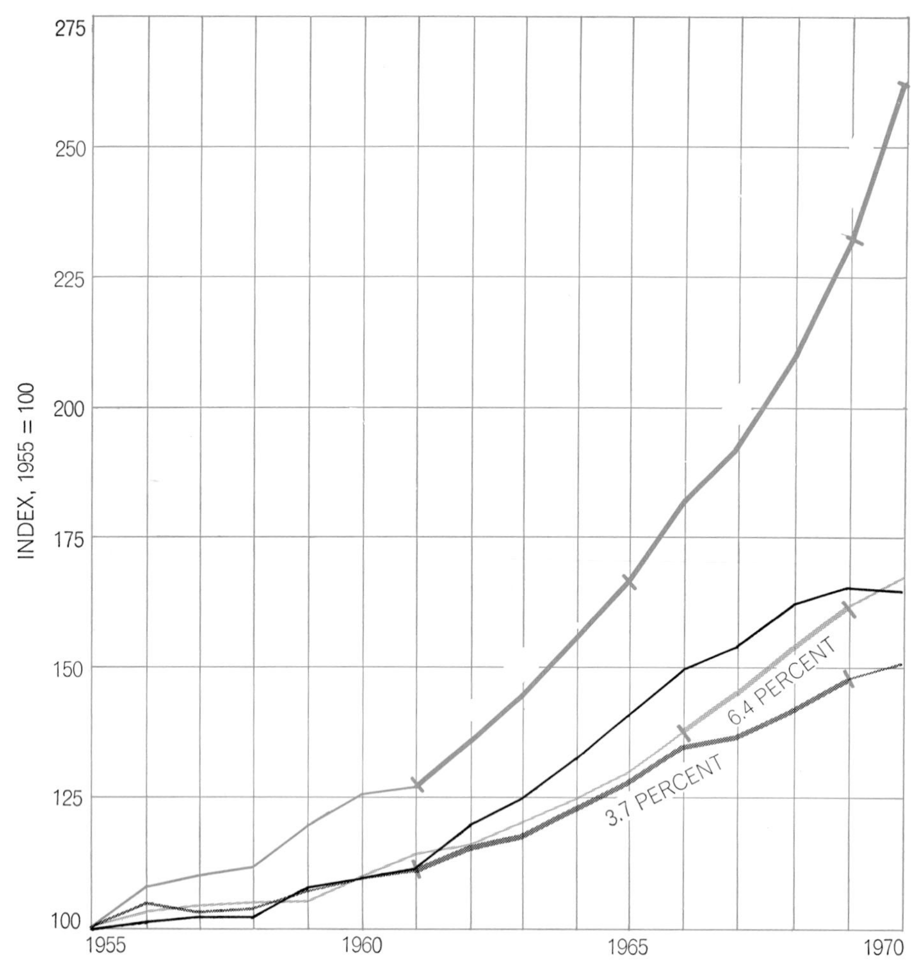

INCREASE IN CONSUMPTION of energy for electricity generation (*dark color*), transportation (*light color*) and other applications (*gray*) and of the gross national product (*black*) are compared. Annual growth rates for certain periods are shown beside heavy segments of curves. Consumption of electricity has a high growth rate and is increasing.

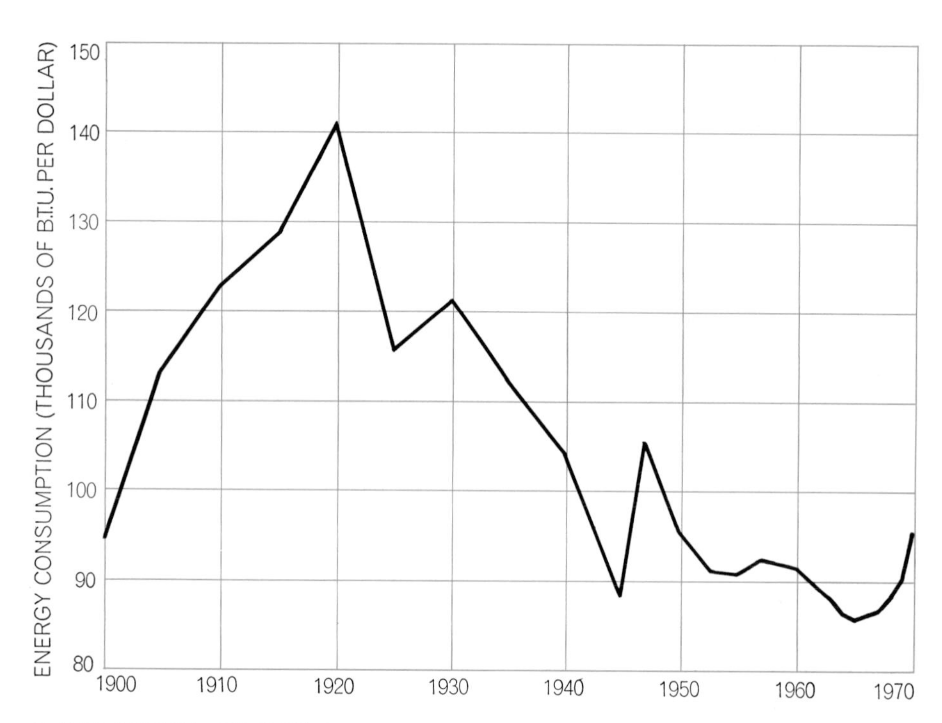

RATIO OF ENERGY CONSUMPTION to gross national product has varied over the years. It tends to be low when the G.N.P. is large and energy is being used efficiently, as was the case during World War II. The ratio has been rising steadily since 1965. Reasons include the increase in the use of air conditioning and the lack of advance in generating efficiency.

sources. The more power an industrial society disposes of, the more it wants. The more power we use, the more we shape our cities and mold our economic and social institutions to be dependent on the application of power and the consumption of energy. We could not now make any major move toward a lower per capita energy consumption without severe economic dislocation, and certainly the struggle of people in less developed regions toward somewhat similar energy-consumption levels cannot be thwarted without prolonging mass human suffering. Yet there is going to have to be some leveling off in the energy demands of industrial societies. Countries such as the U.S. have already come up against constraints dictated by the availability of resources and by damage to the environment. Another article in this issue considers the question of resource availability [see "The Energy Resources of the Earth," by M. King Hubbert, page 31]. Here I shall simply point out some of the decisions the U.S. faces in coping with diminishing supplies, and specifically with our increasing reliance on foreign sources of petroleum and petroleum products. In the short run the advantages of reasonable self-sufficiency must be weighed against the economic and environmental costs of developing oil reserves in Alaska and off the coast of California and the Gulf states. Later on such self-sufficiency may be attainable only through the production of oil from oil shale and from coal. In the long run the danger of dependence on dwindling fossil fuels—whatever they may be —must be balanced against the research and development costs of a major effort to shape a new energy system that is neither dependent on limited resources nor hard on the environment.

The environmental constraint may be more insistent than the constraint of resource availability. The present flow of energy through U.S. society leaves waste rock and acid water at coal mines; spilled oil from offshore wells and tankers; waste gases and particles from power plants, furnaces and automobiles; radioactive wastes of various kinds from nuclear-fuel processing plants and reactors. All along the line waste heat is developed, particularly at the power plants.

Yet for at least the next 50 years we shall be making use of dirty fuels: coal and petroleum. We can improve coal-combustion technology, we can build power plants at the mine mouth (so that the air of Appalachia is polluted instead of the air of New York City), we can make clean oil and gas from coal and oil

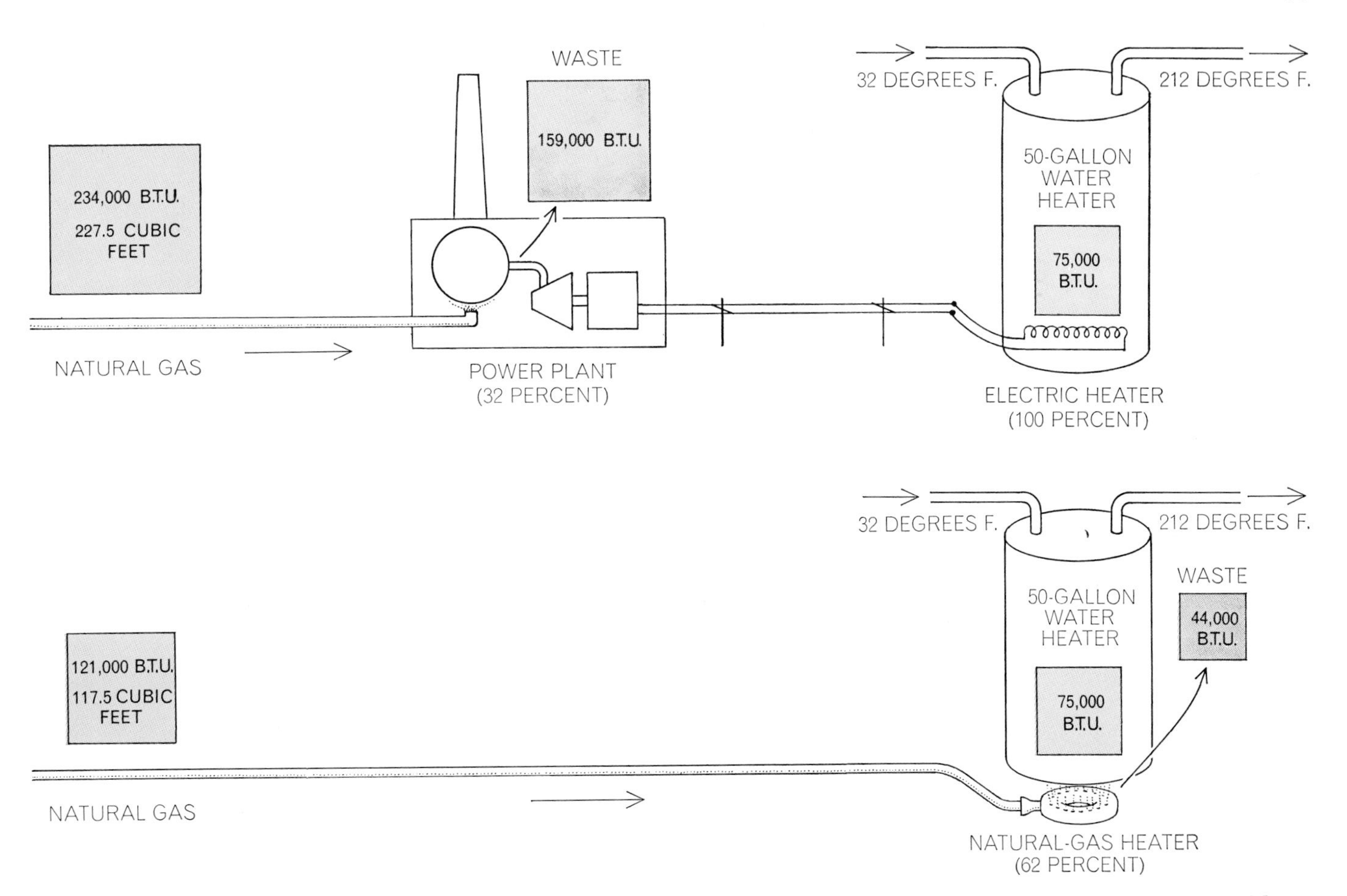

EFFICIENCIES OF HEATING WATER with natural gas indirectly by generating electricity for use in resistance heating (*top*) and directly (*bottom*) are contrasted. In each case the end result is enough heat to warm 50 gallons of water from 32 degrees Fahrenheit to 212 degrees. Electrical method requires substantially more gas even though efficiency at electric heater is nearly 100 percent.

from shale, and sow grass on the mountains of waste. As nuclear power plants proliferate we can put them underground, or far from the cities they serve if we are willing to pay the cost in transmission losses. With adequate foresight, caution and research we may even be able to handle the radioactive-waste problem without "undue" risk.

There are, however, definite limits to such improvements. The automobile engine and its present fuel simply cannot be cleaned up sufficiently to make it an acceptable urban citizen. It seems clear that the internal-combustion engine will be banned from the central city by the year 2000; it should probably be banned right now. Because our cities are shaped for automobiles, not for mass transit, we shall have to develop battery-powered or flywheel-powered cars and taxis for inner-city transport. The 1970 census for the first time showed more metropolitan citizens living in suburbs than in the central city; it also showed a record high in automobiles per capita, with the greatest concentration in the suburbs. It seems reasonable to visualize the suburban two-car garage of the future with one car a recharger for "downtown" and

	PRIMARY ENERGY INPUT (UNITS)	SECONDARY ENERGY OUTPUT (UNITS)	APPLICATION EFFICIENCY (PERCENT)	TECHNICAL EFFICIENCY (PERCENT)
AUTOMOBILE				
INTERNAL-COMBUSTION ENGINE	100		25	25
FLYWHEEL DRIVE CHARGED BY ELECTRICITY	100	32	100	32
SPACE HEATING				
BY DIRECT FUEL USE	100		75	75
BY ELECTRICAL RESISTANCE	100	32	100	32
SMELTING OF STEEL				
WITH COKE	100	94	94	70
WITH ELECTRICITY	100	32	32	32

TECHNICAL EFFICIENCY is the product of conversion efficiency at an intermediate step (if there is one) and application efficiency at the device that does the work. Losses due to friction and heat are ignored in the flywheel-drive automobile data. Coke retains only about 66 percent of the energy of coal, but the energy recovered from the by-products raises the energy conservation to 94 percent.

the other, still gasoline-powered, for suburban and cross-country driving.

Of course, some of the improvement in urban air quality bought by excluding the internal-combustion engine must be paid for by increased pollution from the power plant that supplies the electricity for the nightly recharging of the downtown vehicles. It need not, however, be paid for by an increased draft on the primary energy source; this is one substitution in which electricity need not decrease the technical efficiency of the system. The introduction of heat pumps for space heating and cooling would be another. In fact, the overall efficiency should be somewhat improved and the environmental impact, given adequate attention to the siting, design and operation of the substituting power plant, should be greatly alleviated.

If technology can extend resource availability and keep environmental deterioration within acceptable limits in most respects, the specific environmental problem of waste heat may become the overriding one of the energy system by the turn of the century.

The cooling water required by power plants already constitutes 10 percent of the total U.S. streamflow. The figure will increase sharply as more nuclear plants start up, since present designs of nuclear plants require 50 percent more cooling water than fossil-fueled plants of equal size do. The water is heated 15 degrees Fahrenheit or more as it flows through the plant. For ecological reasons such an increase in water released to a river, lake or ocean bay is unacceptable, at least for large quantities of effluent, and most large plants are now being built with cooling ponds or towers from which much of the heat of the water is dissi-

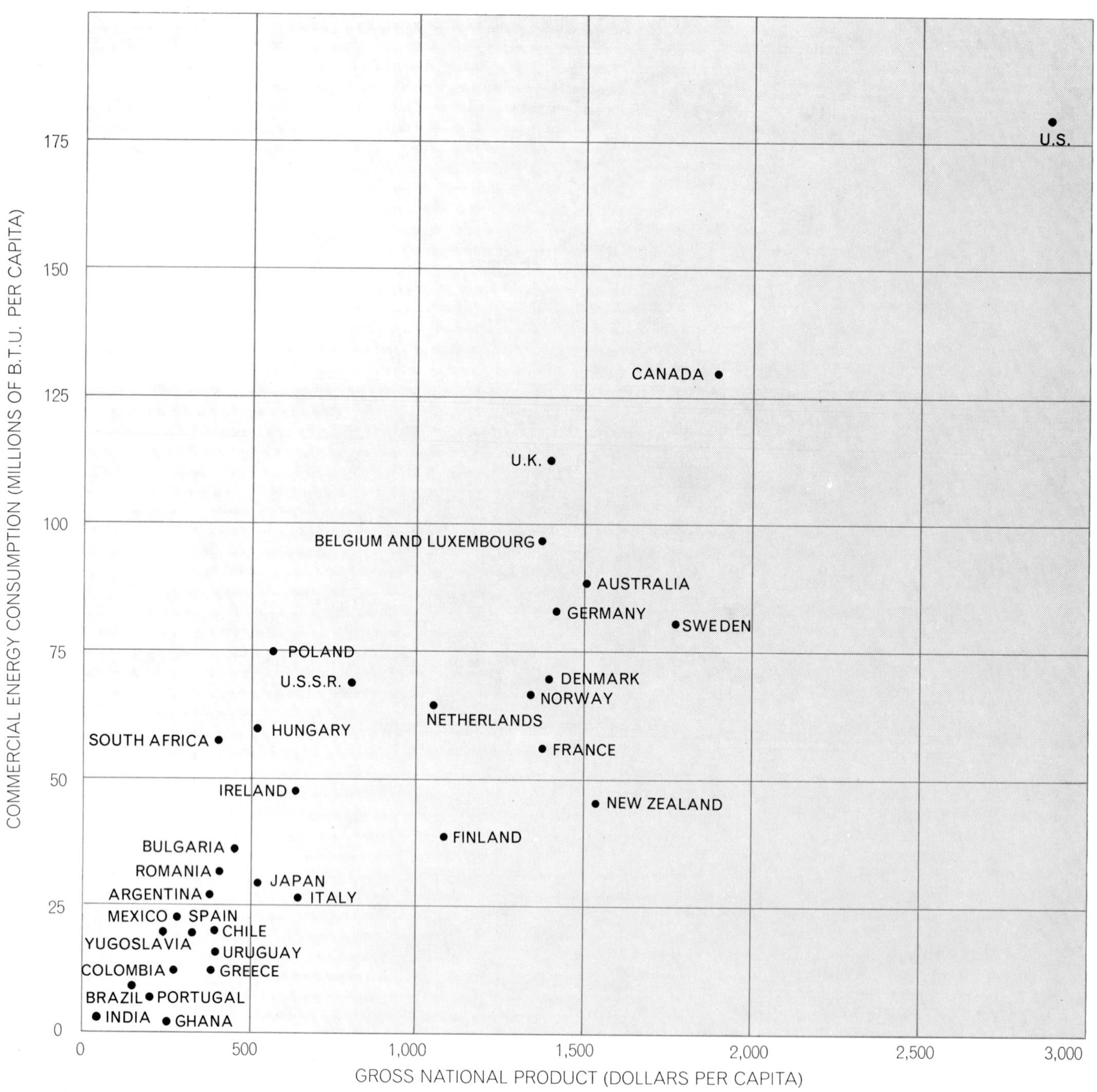

ROUGH CORRELATION between per capita consumption of energy and gross national product is seen when the two are plotted together; in general, high per capita energy consumption is a prerequisite for high output of goods and services. If the position plotted for the U.S. is considered to establish an arbitrary "line," some countries fall above or below that line. This appears to be related to a country's economic level, its emphasis on heavy industry or on services and its efficiency in converting energy into work.

pated to the atmosphere before the water is discharged or recycled through the plant. Although the atmosphere is a more capacious sink for waste heat than any body of water, even this disposal mechanism obviously has its environmental limits.

Many suggestions have been made for putting the waste heat from power plants to work: for irrigation or aquaculture, to provide ice-free shipping lanes or for space heating. (The waste heat from power generation today would be more than enough to heat every home in the U.S.!) Unfortunately the quantities of water involved, the relatively low temperature of the coolant water and the distances between power plants and areas of potential use are serious deterrents to the utilization of waste heat. Plants can be designed, however, for both power production and space heating. Such a plant has been in operation in Berlin for a number of years and has proved to be more efficient than a combination of separate systems for power production and space heating. The Berlin plant is not simply a conserver of waste heat but an exercise in fuel economy; its power capacity was reduced in order to raise the temperature of the heated water above that of normal cooling water.

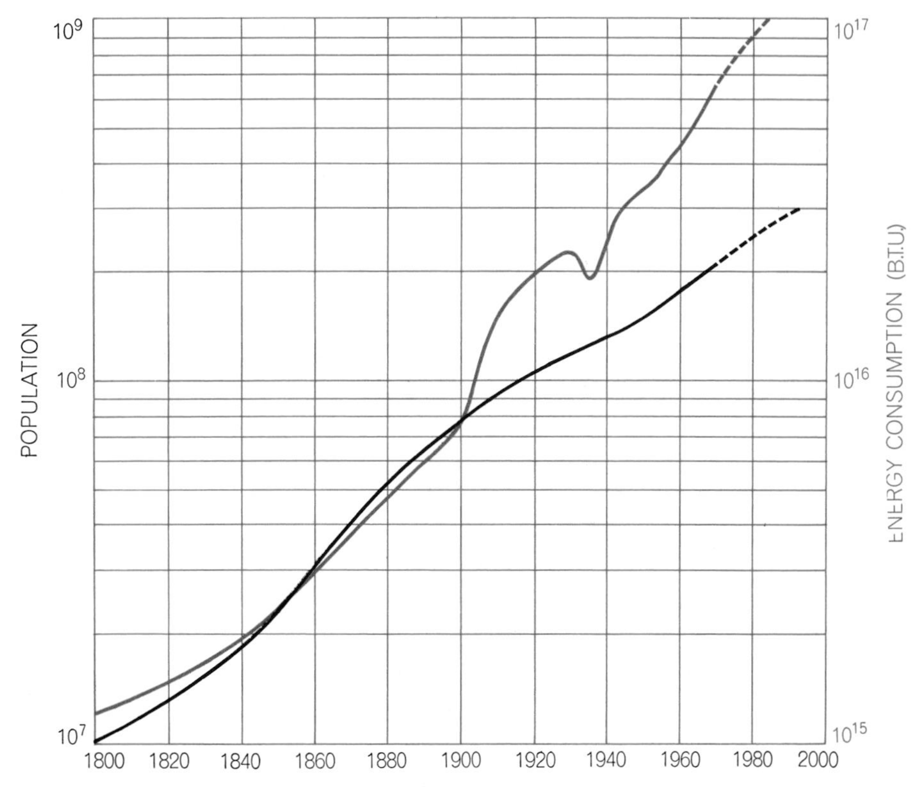

U.S. ENERGY-CONSUMPTION GROWTH (*curve in color*) has outpaced the growth in population (*black*) since 1900, except during the energy cutback of the depression years.

With present and foreseeable technology there is not much hope of decreasing the amount of heat rejected to streams or the atmosphere (or both) from central steam-generating power plants. Two systems of producing power without steam generation offer some long-range hope of alleviating the waste-heat problem. One is the fuel cell; the other is the fusion reactor combined with a system for converting the energy released directly into electricity [see "The Conversion of Energy," by Claude M. Summers, page 95]. In the fuel cell the energy contained in hydrocarbons or hydrogen is released by a controlled oxidation process that produces electricity directly with an efficiency of about 60 percent. A practical fusion reactor with a direct-conversion system is not likely to appear in this century.

Major changes in power technology will be required to reduce pollution and manage wastes, to improve the efficiency of the system and to remove the resource-availability constraint. Making the changes will call for hard political decisions. Energy needs will have to be weighed against environmental and social costs; a decision to set a pollution standard or to ban the internal-combustion engine or to finance nuclear-power development can have major economic and political effects. Democratic societies are not noted for their ability to take the long view in making decisions. Yet indefinite growth in energy consumption, as in human population, is simply not possible.

8

The Conversion of Energy

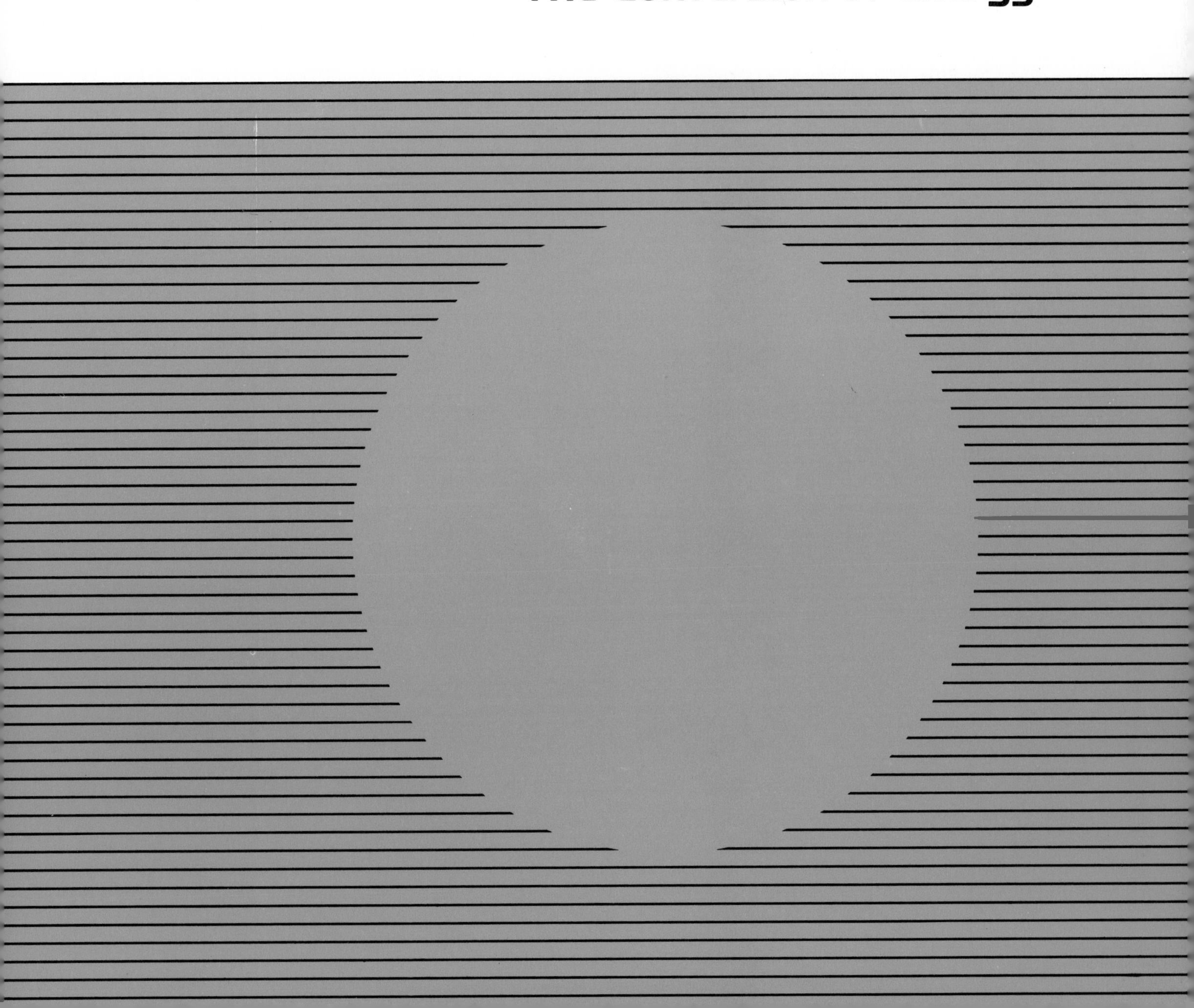

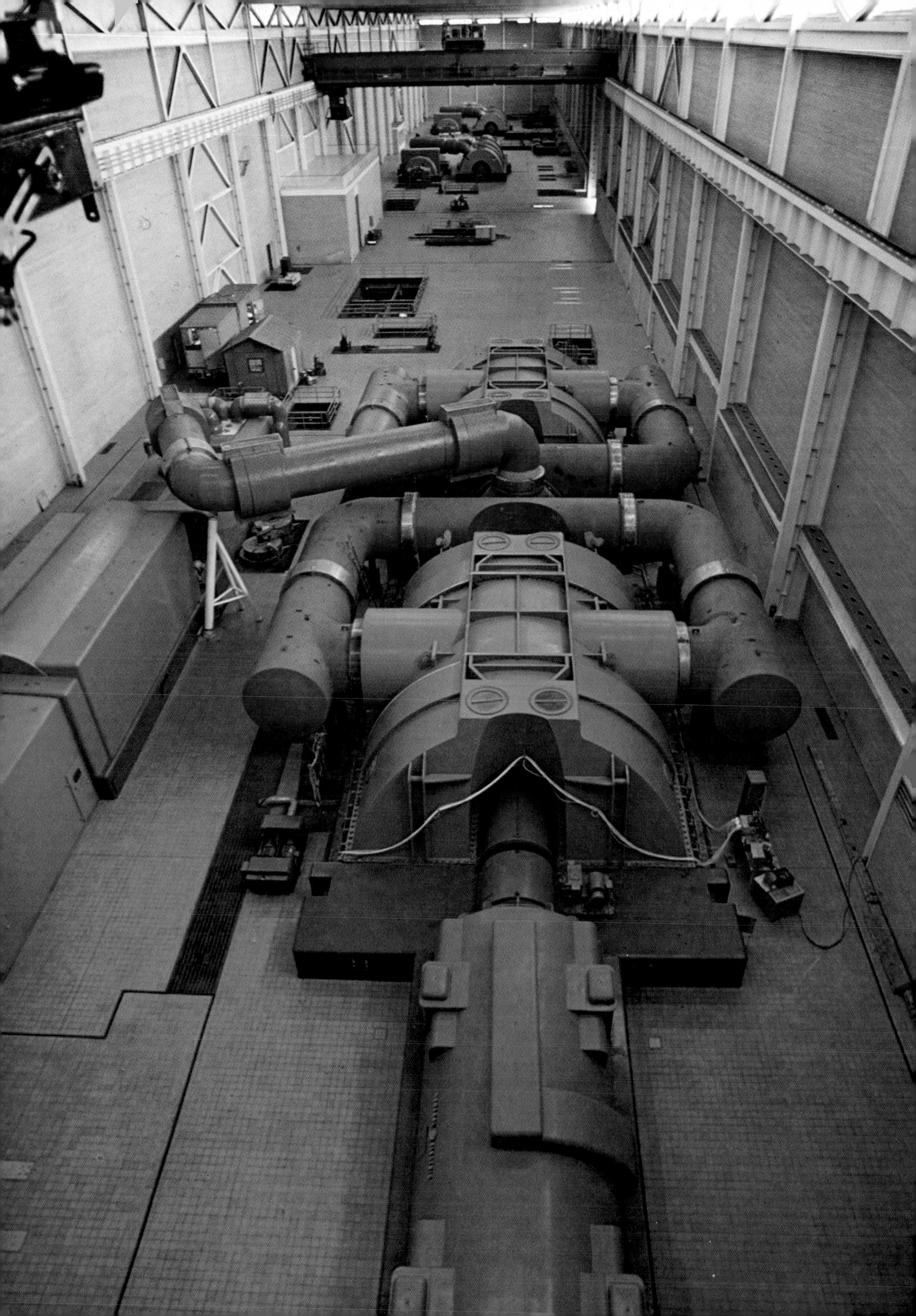

The Conversion of Energy

CLAUDE M. SUMMERS

The efficiency of home furnaces, steam turbines, automobile engines and light bulbs helps in fixing the demand for energy. A major need is a kind of energy source that does not add to the earth's heat load

A modern industrial society can be viewed as a complex machine for degrading high-quality energy into waste heat while extracting the energy needed for creating an enormous catalogue of goods and services. Last year the U.S. achieved a gross national product of just over $1,000 billion with the help of 69×10^{15} British thermal units of energy, of which 95.9 percent was provided by fossil fuels, 3.8 percent by falling water and .3 percent by the fission of uranium 235. The consumption of 340 million B.t.u. per capita was equivalent to the energy contained in about 13 tons of coal or, to use a commodity now more familiar, 2,700 gallons of gasoline. One can estimate very roughly that between 1900 and 1970 the efficiency with which fuels were consumed for all purposes increased by a factor of four. Without this increase the U.S. economy of 1971 would already be consuming energy at the rate projected for the year 2025 or thereabouts.

Because of steadily increasing efficiency in the conversion of energy to useful heat, light and work, the G.N.P. between 1890 and 1960 was enabled to grow at an average annual rate of 3.25 percent while fuel consumption increased at an annual rate of only 2.7 percent. It now appears, however, that this favorable ratio no longer holds. Since 1967 annual increases in fuel consumption have risen faster than the G.N.P., indicating that gains in fuel economy are becoming hard to achieve and that new goods and services are requiring a larger energy input, dollar for dollar, than those of the past. If one considers only the predicted increase in the use of nuclear fuels for generating electricity, it is apparent that an important fraction of the fuel consumed in the 1980's and 1990's will be converted to a useful form at lower efficiency than fossil fuels are today. The reason is that present nuclear plants convert only about 30 percent of the energy in the fuel to electricity compared with about 40 percent for the best fossil-fuel plants.

It is understandable that engineers should strive to raise the efficiency with which fuel energy is converted to other and more useful forms. For industry increased efficiency means lower production costs; for the consumer it means lower prices; for everyone it means reduced pollution of air and water. Electric utilities have long known that by lowering the price of energy for bulk users they can encourage consumption. The recent campaign of the utility industry to "save a watt" marks a profound reversal in business philosophy. The difficulty of finding acceptable new sites for power plants underscores the need not only for frugality of use but also for efficiency of use. Having said this, one must emphasize that even large improvements in efficiency can have only a modest effect in extending the life of the earth's supply of fossil and nuclear fuels. I shall develop the point more fully later in this article.

The efficiency with which energy contained in any fuel is converted to useful form varies widely, depending on the method of conversion and the end use desired. When wood or coal is burned in an open fireplace, less than 20 percent of the energy is radiated into the room; the rest escapes up the chimney. A well-designed home furnace, on the other hand, can capture up to 75 percent of the energy in the fuel and make it available for space heating. The average efficiency of the conversion of fossil fuels for space heating is now probably between 50 and 55 percent, or nearly triple what it was at the turn of the century. In 1900 more than half of all the fuel consumed in the U.S. was used for space heating; today less than a third is so used.

The most dramatic increase in fuel-conversion efficiency in this century has been achieved by the electric-power industry. In 1900 less than 5 percent of the energy in the fuel was converted to electricity. Today the average efficiency is around 33 percent. The increase has been achieved largely by increasing the temperature of the steam entering the turbines that turn the electric generators and by building larger generating units [*see illustration on opposite page*]. In 1910 the typical inlet temperature was 500 degrees Fahrenheit; today the latest

STEAM-DRIVEN TURBOGENERATOR at Paradise power plant of the Tennessee Valley Authority near Paradise, Ky., has a capacity of 1,150 megawatts. When placed in operation in February, 1970, it was the largest unit in the world. The turbine, built by the General Electric Company, is a cross-compound design in which steam first enters a high-pressure turbine, below the angled pipe at the left, then flows through the angled pipe to pass in sequence through an intermediate-pressure turbine (*blue casing at rear*) and then through a low-pressure turbine (*blue casing in foreground*). The high-pressure turbine is connected to one generator (*gray housing at left*) and the other two turbine sections to a second generator of the same capacity (*gray housing in foreground*). The entire unit is driven by eight million pounds of steam per hour, which enters the high-pressure turbine at 3,650 pounds per square inch and 1,003 degrees Fahrenheit. The daily coal consumption is 10,572 tons, enough to fill 210 railroad coal cars. The unit has a net thermal efficiency of 39.3 percent. Two smaller turbogenerators, each rated at 704 megawatts, are visible in the background.

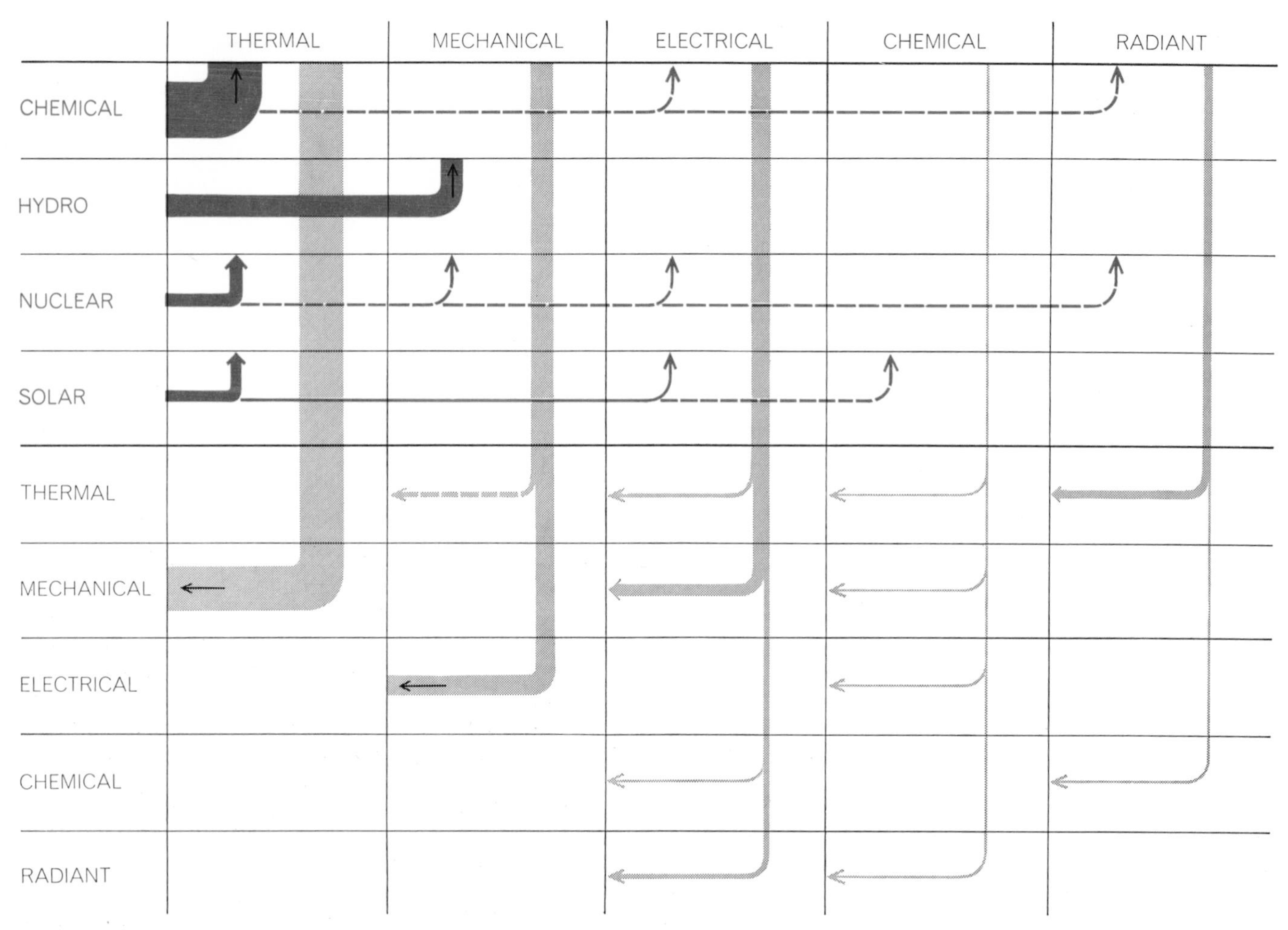

CONVERSION PATHWAYS link many of the familiar forms of energy. The four forms shown in color are either important sources of power today or, in the case of solar energy, potentially important. The broken lines indicate rare, incidental or theoretically useful conversions. The gray lines follow the destiny of intermediate forms of energy. Except for the thermal energy used for space heating, most is converted to mechanical energy. Mechanical energy is used directly for transportation (*see illustration below*) and for generating electricity. Electrical energy in turn is used for lighting, heating and mechanical work. As a secondary form, chemical energy is found in dry cells and storage batteries. The radiant energy produced by electric lamps ends up chiefly as heat.

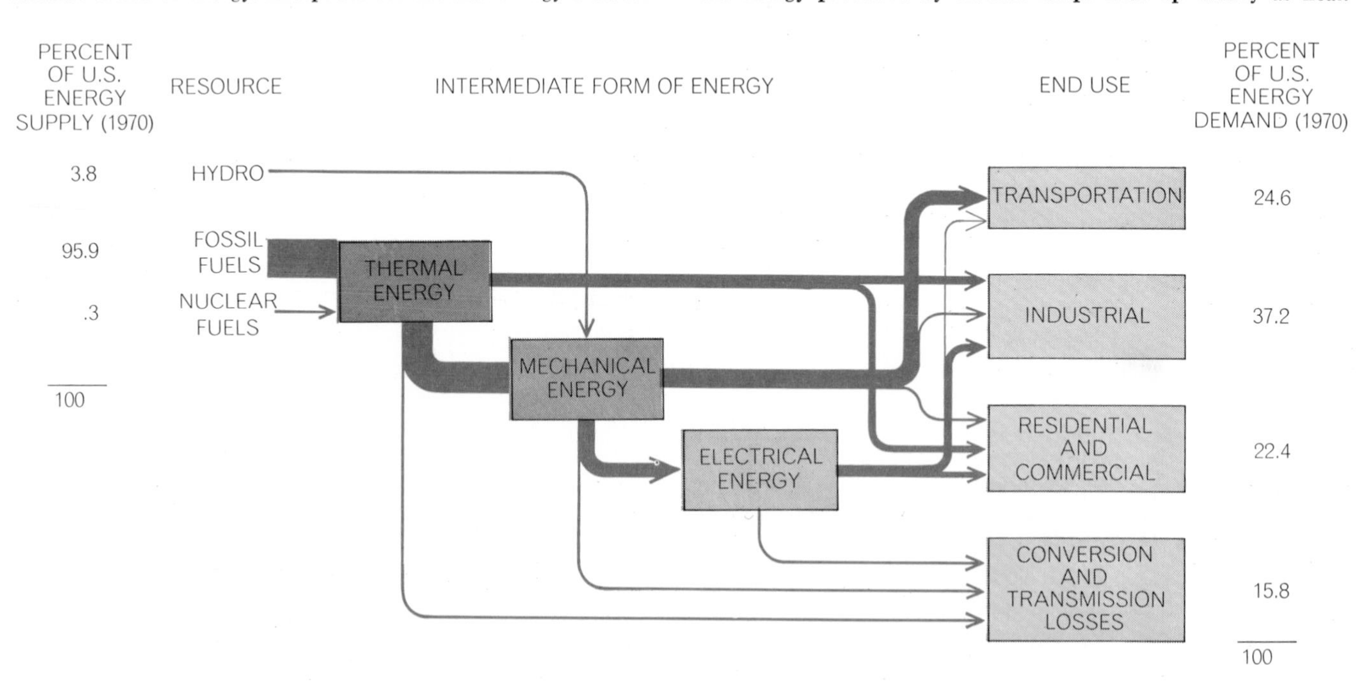

PATHWAYS TO END USES are depicted for the three principal sources of energy. The most direct and most efficient conversion is from falling water to mechanical energy to electrical energy. The energy locked in fossil and nuclear fuels must first be released in the form of thermal energy before it can be converted to mechanical energy and then, if it is desired, to electric power. Conversion and transmission losses include various nonenergy uses of fossil fuels, such as the manufacture of lubricants and the conversion of coal to coke. The biggest loss, however, arises from the generation of electric power at an average efficiency of 32.5 percent.

units take steam superheated to 1,000 degrees. The method of computing the maximum theoretical efficiency of a steam turbine or other heat engine was enunciated by Nicolas Léonard Sadi Carnot in 1824. The maximum achievable efficiency is expressed by the fraction $(T_1 - T_2)/T_1$, where T_1 is the absolute temperature of the working fluid entering the heat engine and T_2 is the temperature of the fluid leaving the engine. These temperatures are usually expressed in degrees Kelvin, equal to degrees Celsius plus 273, which is the difference between absolute zero and zero degrees C. In a modern steam turbine T_1 is typically 811 degrees K. (1,000 degrees Fahrenheit) and T_2 degrees K. (100 degrees F.). Therefore according to Carnot's equation the maximum theoretical efficiency is about 60 percent. Because the inherent properties of a steam cycle do not allow the heat to be introduced at a constant upper temperature, the maximum theoretical efficiency is not 60 percent but more like 53 percent. Modern steam turbines achieve about 89 percent of that value, or 47 percent net.

To obtain the overall efficiency of a steam power plant this value must be multiplied by the efficiencies of the other energy converters in the chain from fuel to electricity. Modern boilers can convert about 88 percent of the chemical energy in the fuel into heat. Generators can convert up to 99 percent of the mechanical energy produced by the steam turbine into electricity. Thus the overall efficiency is 88×47 (for the turbine) $\times$ 99, or about 41 percent.

Nuclear power plants operate at lower efficiency because present nuclear reactors cannot be run as hot as boilers burning fossil fuel. The temperature of the steam produced by a boiling-water reactor is around 350 degrees C., which means that the T_1 in the Carnot equation is 623 degrees K. For the complete cycle from fuel to electricity the efficiency of a nuclear power plant drops to about 30 percent. This means that some 70 percent of the energy in the fuel used by a nuclear plant appears as waste heat, which is released either into an adjacent body of water or, if cooling towers are used, into the surrounding air. For a fossil-fuel plant the heat wasted in this way amounts to about 60 percent of the energy in the fuel.

The actual heat load placed on the water or air is much greater, however, than the difference between 60 and 70 percent suggests. For plants with the same kilowatt rating, a nuclear plant produces about 50 percent more waste

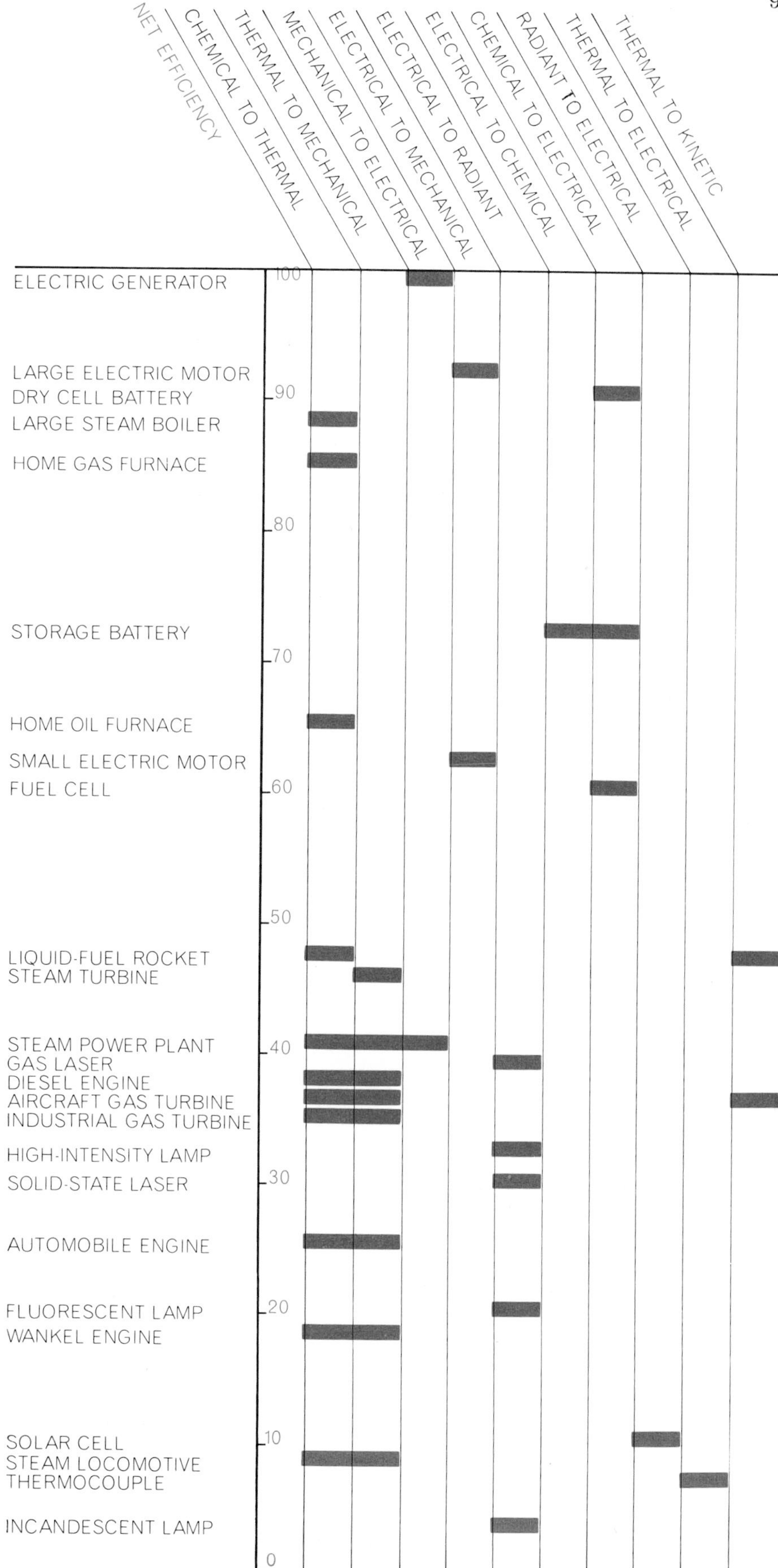

EFFICIENCY OF ENERGY CONVERTERS runs from less than 5 percent for the ordinary incandescent lamp to 99 percent for large electric generators. The efficiencies shown are approximately the best values attainable with present technology. The figure of 47 percent indicated for the liquid-fuel rocket is computed for the liquid-hydrogen engines used in the Saturn moon vehicle. The efficiencies for fluorescent and incandescent lamps assume that the maximum attainable efficiency for an acceptable white light is about 400 lumens per watt rather than the theoretical value of 220 lumens per watt for a perfectly "flat" white light.

heat than a fossil-fuel plant. The reason is that a nuclear plant must "burn" about a third more fuel than a fossil-fuel plant to produce a kilowatt-hour of electricity and then wastes 70 percent of the larger B.t.u. input.

Of course, no law of thermodynamics decrees that the heat released by either a nuclear or a fossil-fuel plant must go to waste. The problem is to find something useful to do with large volumes of low-grade energy. Many uses have been proposed, but all run up against economic limitations. For example, the low-pressure steam discharged from a steam turbine could be used for space heating. This is done in some communities, notably in New York City, where Consolidated Edison is a large steam supplier. Many chemical plants and refineries also use low-pressure steam from turbines as process steam. It has been suggested that the heated water released by power plants might be beneficial in speeding the growth of fish and shellfish in certain localities. Nationwide, however, there seems to be no attractive use for the waste heat from the present fossil-fuel plants or for the heat that will soon be pouring from dozens of new nuclear power plants. The problem will be to limit the harm the heat can do to the environment.

From the foregoing discussion one can see that the use of electricity for home heating (a use that is still vigorously promoted by some utilities) represents an inefficient use of chemical fuel. A good oil- or gas-burning home furnace is at least twice as efficient as the average electric-generating station. In some locations, however, the annual cost of electric space heating is competitive with direct heating with fossil fuels even at the lower efficiency. Several factors account for this anomaly. The electric-power rate decreases with the added load. Electric heat is usually installed in new constructions that are well insulated. The availability of gas is limited in some locations and its cost is higher. The delivery of oil is not always dependable. As fossil fuels become scarcer, their cost will increase, and the production of electrical energy with nuclear fuels will increase. Unfortunately we must expect that a greater percentage of our fuel resources (particularly nuclear fuels) will be consumed in electric space heating in spite of the less efficient use of fuel.

The most ubiquitous of all prime movers is the piston engine. There are two in many American garages, not counting the engines in the power mower, the snowblower or the chain saw. The piston engines in the nation's more than 100 million motor vehicles have a rated capacity in excess of 17 billion horsepower, or more than 95 percent of the capacity of all prime movers (defined as engines for converting fuel to mechanical energy). Although this huge capacity is unemployed most of the time, it accounts for more than 16 percent of the fossil fuel consumed by the U.S. Transportation in all forms—including the propulsion systems of ships, locomotives and aircraft—absorbs about 25 percent of the nation's energy budget.

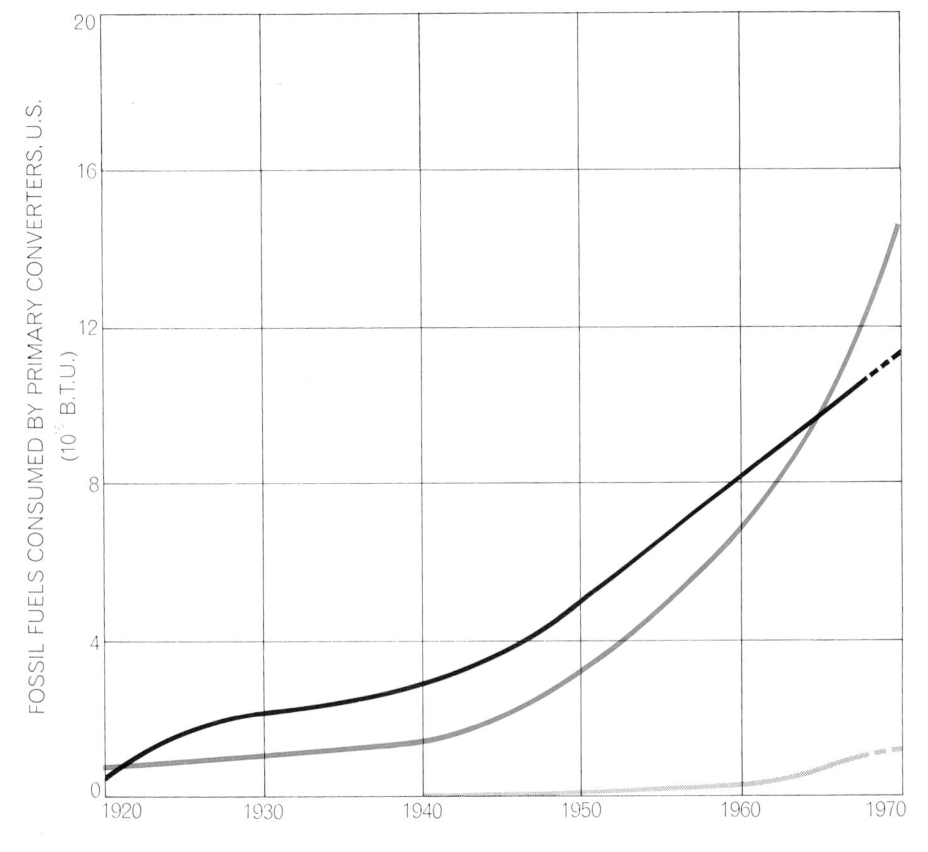

THREE OF FASTEST-GROWING ENERGY USERS are electric utilities (*color*), motor vehicles (*black*) and aircraft (*gray*). Together they now consume about 40 percent of all the energy used in the U.S. As recently as 1940 they accounted for only 18 percent of a much smaller total. The demand for aircraft fuel has more than tripled in 10 years.

Automotive engineers estimate that the efficiency of the average automobile engine has risen about 10 percent over the past 50 years, from roughly 22 percent to 25 percent. In terms of miles delivered per gallon of fuel, however, there has actually been a decline. From 1920 until World War II the average automobile traveled about 13.5 miles per gallon of fuel. In the past 25 years the average has fallen gradually to about 12.2 miles per gallon. This decline is due to heavier automobiles with more powerful engines that encourage greater acceleration and higher speed. It takes about eight times more energy to push a vehicle through the air at 60 miles per hour than at 30 miles per hour. The same amount of energy used in accelerating the car's mass to 60 miles per hour must be absorbed as heat, primarily in the brakes, to stop the vehicle. Therefore most of the gain in engine efficiency is lost in the way man uses his machine. Automobile air conditioning has also played a role in reducing the miles per gallon. With the shift in consumer preference to smaller cars the figure may soon begin to climb. The efficiency of the basic piston engine, however, cannot be improved much further.

If all cars in the year 2000 operated on electric batteries charged by electricity generated in central power stations, there would be little change in the nation's overall fuel requirement. Although the initial conversion efficiency in the central station might be 35 percent compared with 25 percent in the piston engine, there would be losses in distributing the electrical energy and in the conversion of electrical energy to chemical energy (in the battery) and back to electrical energy to turn the car wheels. Present storage batteries have an overall efficiency of 70 to 75 percent, so that there is not much room for improvement. Anyone who believes we would all be better off if cars were electrically powered must consider the problem of increasing the country's electric-generating capacity by about 75 percent, which is what would be required to move 100 million vehicles.

The difficulty of trying to trace savings produced by even large changes in efficiency of energy conversion is vividly demonstrated by what happened when the railroads converted from the steam locomotive (maximum thermal efficiency 10 percent) to diesel-electric locomotives (thermal efficiency about 35 percent). In 1920 the railroads used about 135 million tons of coal, which represented 16 percent of the nation's total energy demand. By 1967, according to estimates made by John Hume, an energy consultant, the railroads were providing 54 percent more transportation than in 1920 (measured by an index of "transportation output") with less than a sixth as many B.t.u. This increase in efficiency, together with the railroads' declining role in the national economy, had reduced the railroads' share of the nation's total fuel budget from 16 percent to about 1 percent. If one looks at a curve of the country's total fuel consumption from 1920 to 1967, however, the impact of this extraordinary change is scarcely visible.

Perhaps the least efficient important use for electricity is providing light. The General Electric Company estimates that lighting consumes about 24 percent of all electrical energy generated, or 6 percent of the nation's total energy budget. It is well known that the glowing filament of an ordinary 100-watt incandescent lamp produces far more heat than light. In fact, more than 95 percent of the electric input emerges as infrared radiation and less than 5 percent as visible light. Nevertheless, this is about five times more light than was provided by a 100-watt lamp in 1900. A modern fluorescent lamp converts about 20 percent of the electricity it consumes into light. These values are based on a practical upper limit of 400 lumens per watt, assuming the goal is an acceptable light of less than perfect whiteness. If white light with a totally flat spectrum is specified, the maximum theoretical output is reduced to 220 lumens per watt. If one were satisfied with light of a single wavelength at the peak sensitivity of the human eye (555 nanometers), one could theoretically get 680 lumens per watt.

General Electric estimates that fluorescent lamps now provide about 70 percent of the country's total illumination and that the balance is divided between incandescent lamps and high-

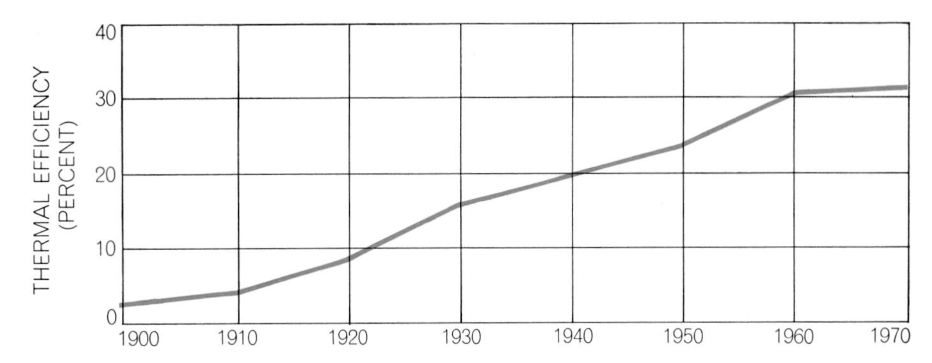

EFFICIENCY OF FUEL-BURNING POWER PLANTS in the U.S. increased nearly tenfold from 3.6 percent in 1900 to 32.5 percent last year. The increase was made possible by raising the operating temperature of steam turbines and increasing the size of generating units.

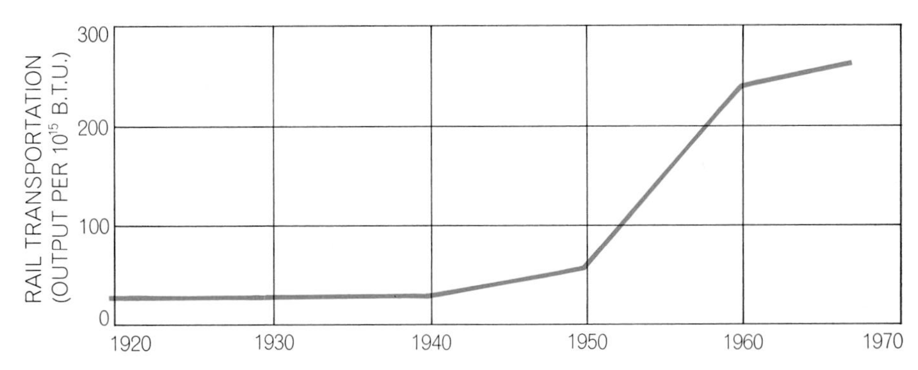

EFFICIENCY OF RAILROAD LOCOMOTIVES can be inferred from the energy needed by U.S. railroads to produce a unit of "transportation output." The big leap in the 1950's reflects the nearly complete replacement of steam locomotives by diesel-electric units.

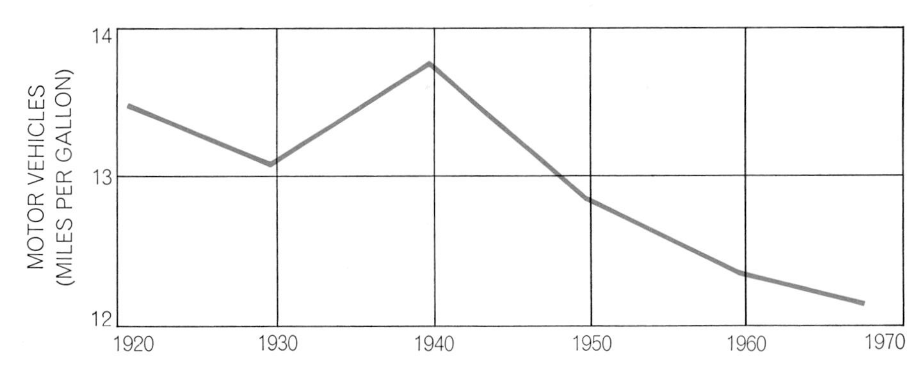

EFFICIENCY OF AUTOMOBILE ENGINES is reflected imperfectly by miles per gallon of fuel because of the increasing weight and speed of motor vehicles. Manufacturers say that the thermal efficiency of the 1920 engine was about 22 percent; today it is about 25 percent.

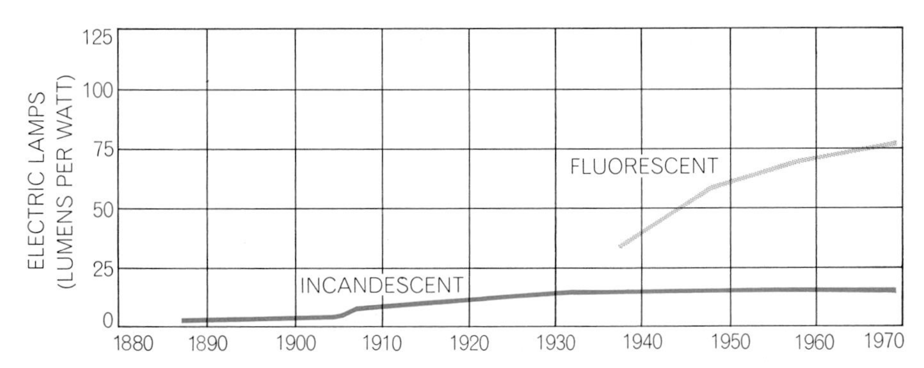

EFFICIENCY OF ELECTRIC LAMPS depends on the quality of light one regards as acceptable. Theoretical efficiency for perfectly flat white light is 220 lumens per watt. By enriching the light slightly in mid-spectrum one could obtain about 400 lumens per watt. Thus present fluorescent lamps can be said to have an efficiency of either 36 percent or 20.

intensity lamps, which have efficiencies comparable to, and in some cases higher than, fluorescent lamps. This division implies that the average efficiency of converting electricity to light is about 13 percent. To obtain an overall efficiency for converting chemical (or nuclear) energy to visible light, one must multiply this percentage times the average efficiency of generating power (33 percent), which yields a net conversion efficiency of roughly 4 percent. Nevertheless, thanks to increased use of fluorescent and high-intensity lamps, the nation was able to triple its "consumption" of lighting between 1960 and 1970 while only doubling the consumption of electricity needed to produce it.

This brief review of changing efficiencies of energy use may provide some perspective when one tries to evaluate

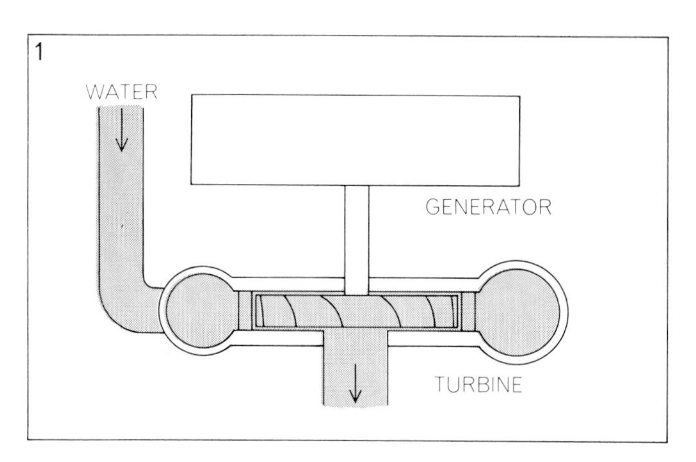

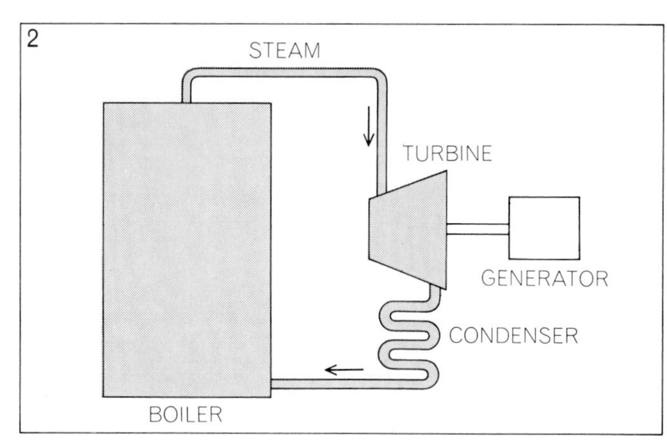

ELECTRIC-POWER GENERATING MACHINERY now in use extracts energy from falling water, fossil fuels or nuclear fuels. The hydroturbine generator (*1*) converts potential and kinetic energy into electric power. In a fossil-fuel steam power plant (*2*) a boiler produces steam; the steam turns a turbine; the turbine turns an electric generator. In a nuclear power plant (*3*) the fission of ura-

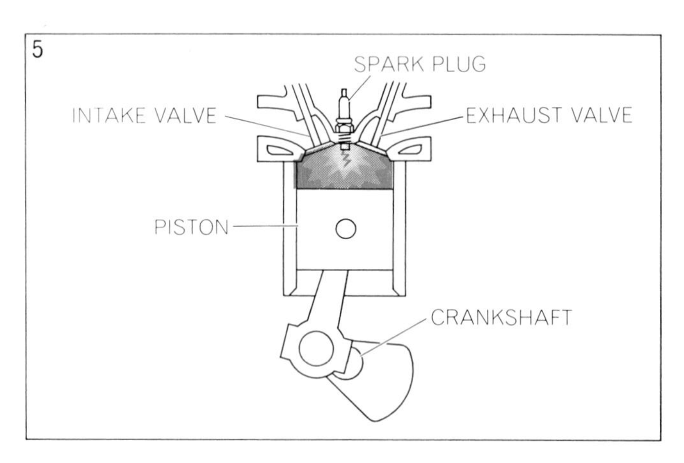

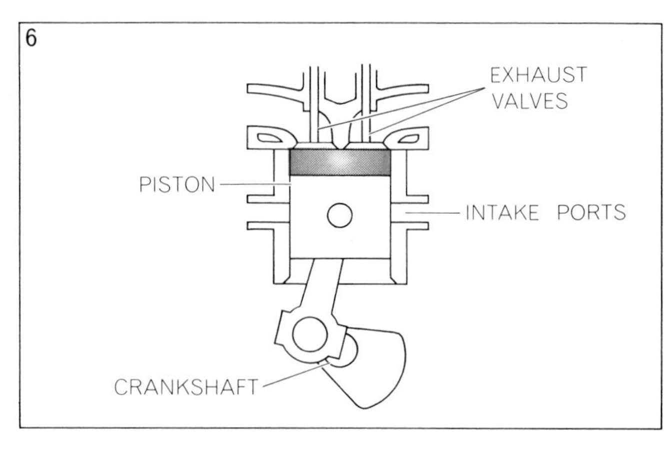

PROPULSION MACHINERY converts the energy in liquid fuels into forms of mechanical or kinetic energy useful for work and transportation. In the piston engine (*5*) a compressed charge of fuel and air is exploded by a spark; the expanding gases push against the piston, which is connected to a crankshaft. In a diesel engine (*6*) the compression alone is sufficient to ignite the charge

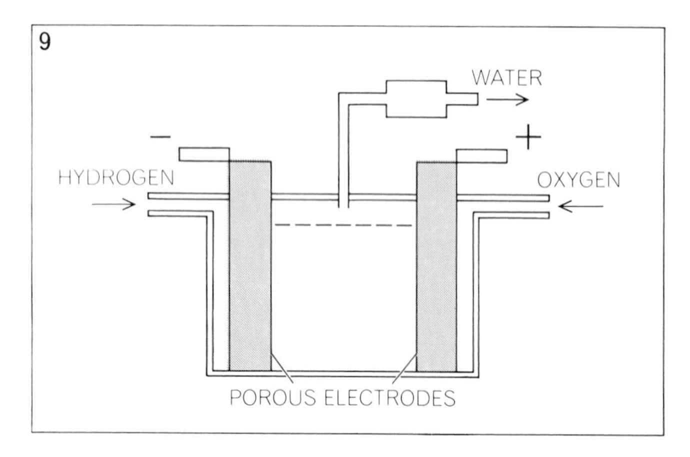

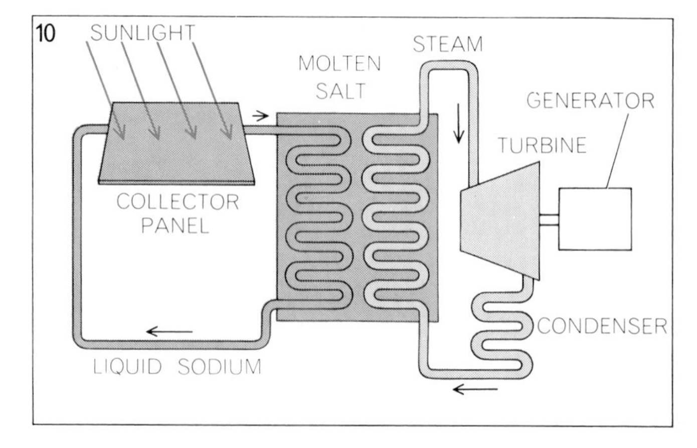

NOVEL ENERGY CONVERTERS are being designed to exploit a variety of energy sources. The fuel cell (*9*) converts the energy in hydrogen or liquid fuels directly into electricity. The "combustion" of the fuel takes place inside porous electrodes. In a recently proposed solar power plant (*10*) sunlight falls on specially coated collectors and raises the temperature of a liquid metal to 1,000 degrees F. A heat exchanger transfers the heat so collected to steam, which then turns a turbogenerator as in a conventional power plant. A salt reservoir holds enough heat to keep generating steam during the night and when the sun is hidden by clouds. In a mag-

the probable impact of novel energy-conversion systems now under development. Two devices that have received much notice are the fuel cell and the magnetohydrodynamic (MHD) generator. The former converts chemical energy directly into electricity; the latter is potentially capable of serving as a high-temperature "topping" device to be operated in series with a steam turbine and generator in producing electricity. Fuel cells have been developed that can "burn" hydrogen, hydrocarbons or alcohols with an efficiency of 50 to 60 percent. The hydrogen-oxygen fuel cells used in the Apollo space missions, built by the Pratt & Whitney division of United Aircraft, have an output of 2.3 kilowatts of direct current at 20.5 volts.

A decade ago the magnetohydrodynamic generator was being advanced as

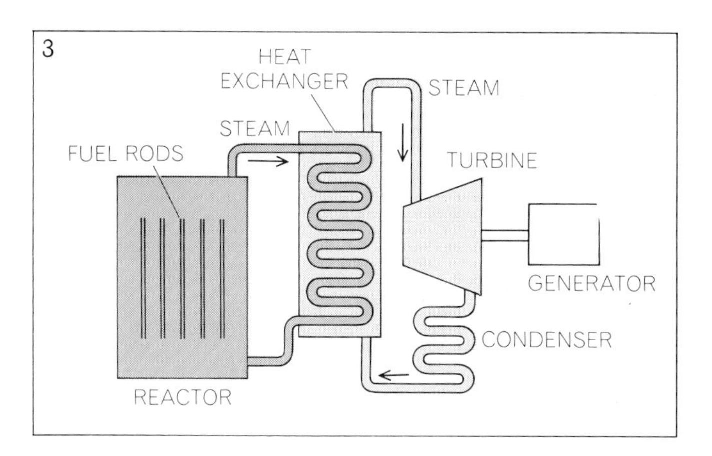

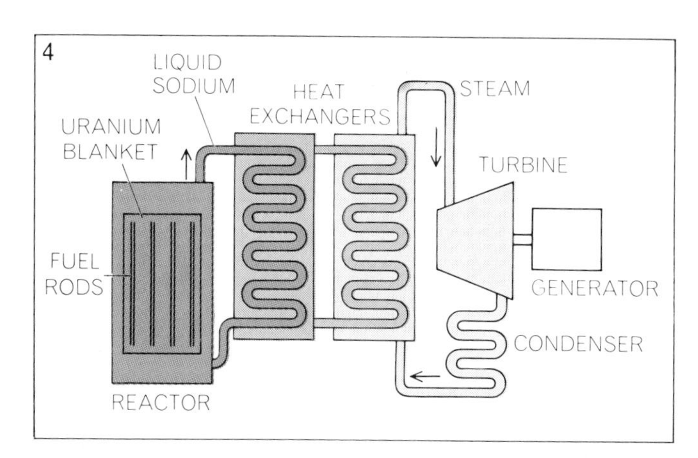

nium 235 releases the energy to make steam, which then goes through the same cycle as in a fossil-fuel power plant. Under development are nuclear breeder reactors (*4*) in which surplus neutrons are captured by a blanket of nonfissile atoms of uranium 238 or thorium 232, which are transformed into fissile plutonium 239 or U-233. The heat of the reactor is removed by liquid sodium.

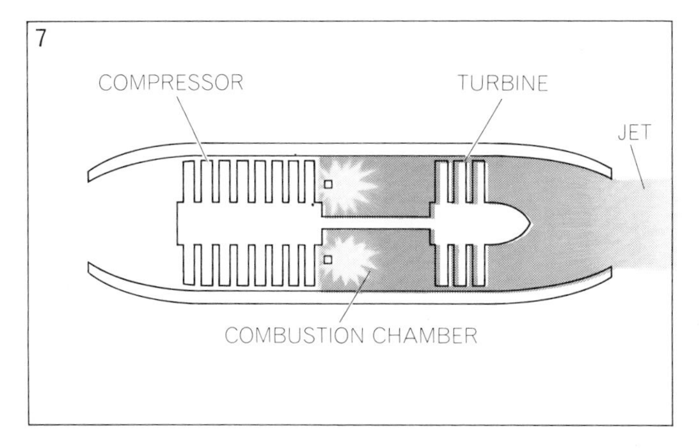

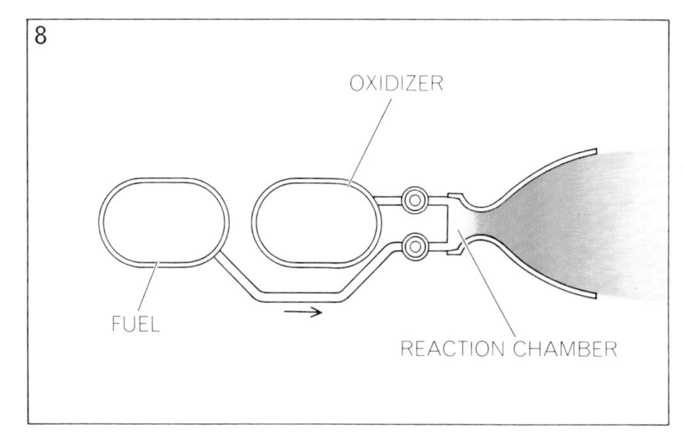

of fuel and air. In an aircraft gas turbine (*7*) the continuous expansion of hot gas from the combustion chamber passes through a turbine that turns a multistage air compressor. Hot gases leaving the turbine provide the kinetic energy for propulsion. A liquid-fuel rocket (*8*) carries an oxidizer in addition to fuel so that it is independent of an air supply. Rocket exhaust carries kinetic energy.

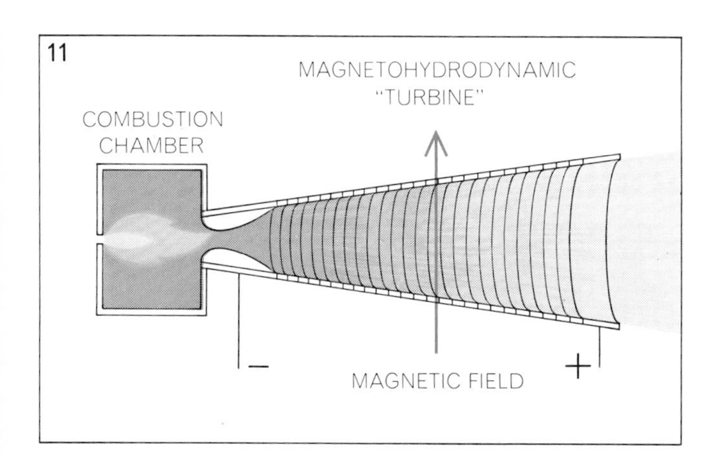

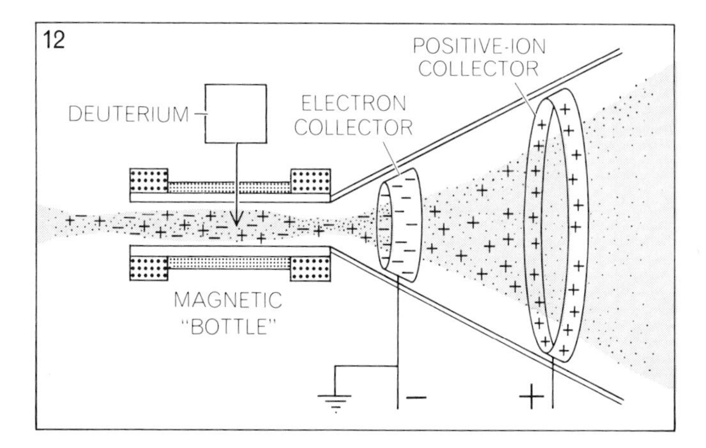

netohydrodynamic "turbine" (*11*) the energy contained in a hot electrically conducting gas is converted directly into electric power. A small amount of "seed" material, such as potassium carbonate, must be injected into the flame to make the hot gas a good conductor. Electricity is generated when the electrically charged particles of gas cut through the field of an external magnet. A long-range goal is a thermonuclear reactor (*12*) in which the nuclei of light elements fuse into heavier elements with the release of energy. High-velocity charged particles produced by a thermonuclear reaction might be trapped in such a way as to generate electricity directly.

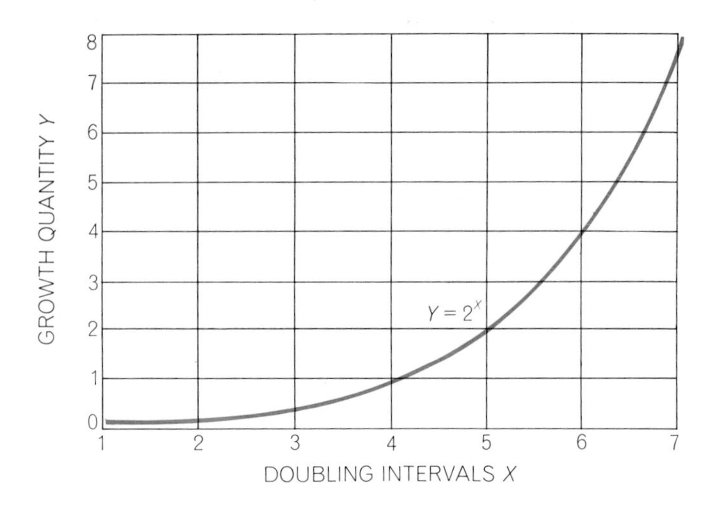

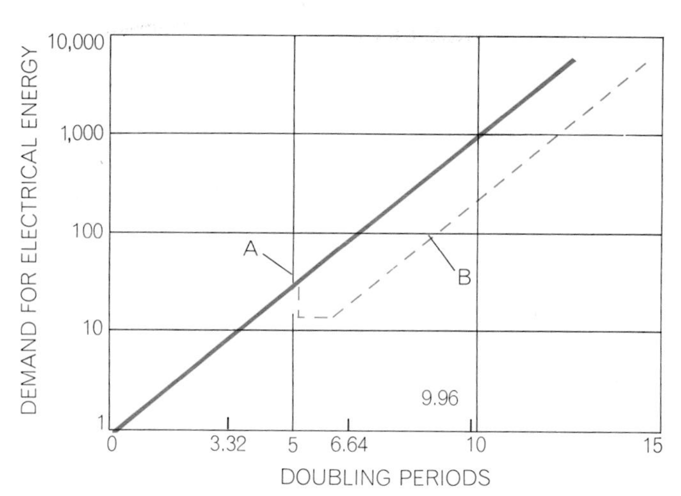

DOUBLING CURVE (*left*) **rises exponentially with time. It shows how many doubling intervals are needed to produce a given multiplication of the growth quantity. Thus if electric-power demand continues to double every 10 years, the demand will increase eightfold in three doubling periods, that is, by the year 2001. When exponential growth curves are plotted on a semilogarithmic scale, the result is a straight line (*right*). If electric-power consumption were cut in half at *A*, held constant for 10 years and allowed to return to the former growth rate, time needed to reach a given demand (*B*) would be extended by only two doubling periods, or 20 years.**

SATURATION OR DEPLETION (PERCENT)	DOUBLING PERIODS TO REACH 100 PERCENT	YEARS FROM NOW
100.0	0.0	0
10.0	3.32	33
1.0	6.64	66
.1	9.96	100
.01	13.28	133
.001	16.60	166
.0001	19.92	199
.00001	23.24	232
.000001	26.56	266
.0000001	29.88	299
.00000001	33.20	332

DEPLETION OF A RESOURCE can be read from the curve at the left. Thus if .1 percent of world's oil has now been extracted, all will be gone in just under 10 doubling periods, or 100 years if the doubling interval is 10 years. The table at the right shows that the ultimate depletion date is changed very little by large changes in the estimate of amount of resource that has been extracted to date.

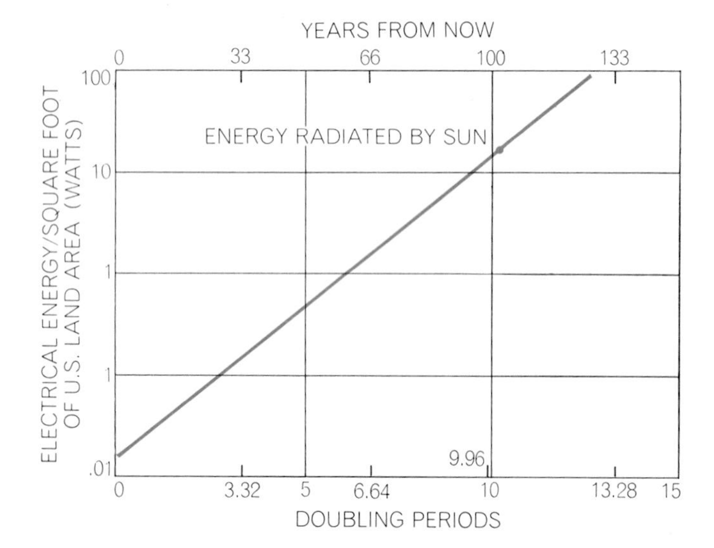

ENERGY RECEIVED FROM SUN on an average square foot of the U.S. will be equaled by production of electrical energy in roughly 100 years if the demand continues to double every 10 years.

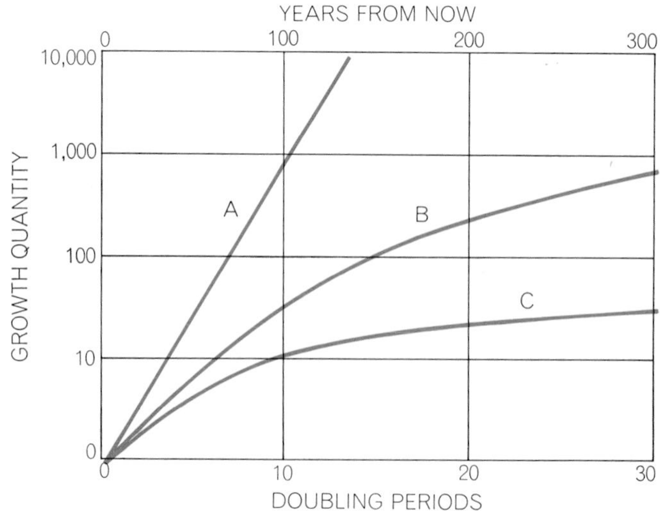

THREE GROWTH CURVES are compared. Curve *A* is exponential. In Curve *B* each doubling period is successively increased by 20 percent. In Curve *C* the growth per doubling period is constant.

the energy converter of the future. In such a device the fuel is burned at a high temperature and the gaseous products of combustion are made electrically conducting by the injection of a "seed" material, such as potassium carbonate. The electrically conducting gas travels at high velocity through a magnetic field and in the process creates a flow of direct current [*see No. 11 in illustrations on pages 100 and 101*]. If the MHD technology can be developed, it should be possible to design fossil-fuel power plants with an efficiency of 45 to 50 percent. Since MHD requires very high temperatures it is not suitable for use with nuclear-fuel reactors, which produce a working fluid much cooler than one can obtain from a combustion chamber fired with fossil fuel.

If ever an energy source can be said to have arrived in the nick of time, it is nuclear energy. Twenty-two nuclear power plants are now operating in the U.S. Another 55 plants are under construction and more than 40 are on order. This year the U.S. will obtain 1.4 percent of its electrical energy from nuclear fission; it is expected that by 1980 the figure will reach 25 percent and that by 2000 it will be 50 percent.

Although a 1,000-megawatt nuclear power plant costs about 10 percent more than a fossil-fuel plant ($280 million as against $250 million), nuclear fuel is already cheaper than coal at the mine mouth. Some projections indicate that coal may double in price between now and 1980. One reason given is that new Federal safety regulations have already reduced the number of tons produced per man-day from the 20 achieved in 1969 to fewer than 15.

The utilities are entering a new period in which they will have to rethink the way in which they meet their base load, their intermediate load (which coincides with the load added roughly between 7:00 A.M. and midnight by the activity of people at home and at work) and peak load (the temperature-sensitive load, which accounts for only a few percent of the total demand). In the past utilities assigned their newest and most efficient units to the base load and called on their older and smaller units to meet the variable daily demand. In the future, however, when still newer capacity is added, the units now carrying the basic load cannot easily be relegated to intermittent duty because they are too large to be easily put on the line and taken off.

There is therefore a need for a new kind of flexible generating unit that may be best satisfied by coupling an industrial gas turbine to an electric generator and using the waste heat from the gas turbine to produce low-pressure steam for a steam turbine–generator set. Combination systems of this kind are now being offered by General Electric and the Westinghouse Electric Company. Although somewhat less efficient than the best large conventional units, the gas-turbine units can be brought up to full load in an hour and can be installed at lower cost per kilowatt. To meet brief peak demands utilities are turning to gas turbines (without waste-heat boilers that can be brought up to full load in minutes) and to pumped hydrostorage systems. In the latter systems off-peak capacity is used to pump water to an elevated reservoir from which it can be released to produce power as needed.

Westinghouse has recently estimated that U.S. utilities must build more than 1,000 gigawatts (GW, or 10^9 watts) of new capacity between 1970 and 1990, or more than three times the present installed capacity of roughly 300 GW. Of the new capacity 500 GW, or half, will be needed to handle the anticipated increase in base load and 75 percent of the 500 GW will be nuclear. More than 400 GW of new capacity will be needed to meet the growing intermediate load, and a sizable fraction of it will be provided by gas turbines. The new peaking capacity, amounting to some 170 GW, will be divided, Westinghouse believes, between gas turbines and pumped storage in the ratio of 10 to seven.

Such projections can be regarded as the conventional wisdom. Does unconventional wisdom have anything to offer that may influence power generation by 2000, if not by 1990? First of all, there are the optimists who believe prototype nuclear-fusion plants will be built in the 1980's and full-scale plants in the 1990's. In a sense, however, this is merely conventional wisdom on an accelerated time scale. Those with a genuinely unconventional approach are asking: Why do we not start developing the technology to harness energy from the sun or the wind or the tides?

Many people still remember the Passamaquoddy project of the 1930's, which is once more being discussed and which would provide 300 megawatts (less than a third the capacity of the turbogenerator shown on page 2) by exploiting tides with an average range of 18 feet in the Bay of Fundy, between Maine and Canada. A working tidal power plant of 240 megawatts has recently been placed in operation by the French government in the estuary of the Rance River, where the tides average 27 feet. How much tidal energy might the U.S. extract if all favorable bays and inlets were developed? All estimates are subject to heavy qualification, but a reasonable guess is something like 100 GW. We have just seen, however, that the utilities will have to add 10 times that much capacity just to meet the needs of 1990. One must conclude that tidal power does not qualify as a major unconventional resource.

What about the wind? A study we conducted at Oklahoma State University a few years ago showed that the average wind energy in the Oklahoma City area is about 18.5 watts per square foot of area perpendicular to the wind direction. This is roughly equivalent to the amount of solar energy that falls on a square foot of land in Oklahoma, averaging the sunlight for 24 hours a day in all seasons and under all weather conditions. A propeller-driven turbine could convert the wind's energy into electricity at an efficiency of somewhere between 60 and 80 percent. Like tidal energy and other forms of hydropower, wind power would have the great advantage of not introducing waste heat into the biosphere.

The difficulty of harnessing the wind's energy comes down to a problem of energy storage. Of all natural energy sources the wind is the most variable. One must extract the energy from the wind as it becomes available and store it if one is to have a power plant with a reasonably steady output. Unfortunately technology has not yet produced a practical storage medium. Electric storage batteries are out of the question.

One scheme that seems to offer promise is to use the variable power output of a wind generator to decompose water into hydrogen and oxygen. These would be stored under pressure and recombined in a fuel cell to generate electricity on a steady basis [*see illustration on next page*]. Alternatively the hydrogen could be burned in a gas turbine, which would turn a conventional generator. The Rocketdyne Division of North American Rockwell has seriously proposed that an industrial version of the hydrogen-fueled rocket engine it builds for the Saturn moon vehicle could be used to provide the blast of hot gas needed to power a gas turbine coupled to an electric generator. Rocketdyne visualizes that a water-cooled gas turbine could operate at a higher temperature than conventional fuel-burning gas turbines and achieve our overall plant efficiency of 55 percent. If the Rocketdyne concept were successful, it could use hydrogen from any source. A wind-driven hydrogen-rocket gas-turbine power plant should be unconventional enough to please the most exotic taste.

By comparison most proposals for harnessing solar energy seem tame indeed. One fairly straightforward proposal has recently been made to the Arizona Power Authority on behalf of the University of Arizona by Aden B. Meinel and Marjorie P. Meinel of the university's Optical Sciences Center. They suggest that if the sunlight falling on 14 percent of the western desert regions of the U.S. were efficiently collected, it could be converted into 1,000 GW of power, or approximately the amount of additional power needed between now and 1990. The Meinels believe it is within the reach of present technology to design collecting systems capable of storing solar energy as heat at 1,000 degrees F., which could be converted to electricity at an overall efficiency of 30 percent.

The key to the project lies in recently developed surface coatings that have high absorbance for solar radiation and low emittance in the infrared region of the spectrum. To achieve a round-the-clock power output, heat collected during daylight hours would be stored in molten salts at 1,000 degrees F. A heat exchanger would transfer the stored energy to steam at the same temperature. The thermal storage tank for a 1,000-megawatt generating plant would require a capacity of about 300,000 gallons. The Meinels propose that industry and the Government immediately undertake design and construction of a 100-megawatt demonstration plant in the vicinity of Yuma, Ariz. The collectors for such a plant would cover an area of 3.6 million square meters (slightly more than a square mile). The Meinels estimate that after the necessary development has been done a 1,000-megawatt solar power station might be built for about $1.1 billion, or about four times the present cost of a nuclear power plant. As they point out: "Solar power faces the economic problem that energy is purchased via a capital outlay rather than an operating expense." They calculate nevertheless that a plant with an operating lifetime of 40 years should produce power at an average cost of only half a cent per kilowatt hour.

A more exotic solar-power scheme has been advanced by Peter E. Glaser of Arthur D. Little, Inc. The idea is to place a lightweight panel of solar cells in a synchronous orbit 22,300 miles above the earth, where they would be exposed to sunlight 24 hours a day. Solar cells (still to be developed) would collect the radiant energy and convert it to electricity with an efficiency of 15 to 20 percent. The electricity would then be converted electronically in orbit to microwave energy with an efficiency of 85 percent, which is possible today. The microwave radiation would be at a wavelength selected to penetrate clouds with little or no loss and would be collected by a suitable antenna on the earth. Present techniques can convert microwave energy to electric power with an efficiency of about 70 percent, and 80 to 85 percent should be attainable. Glaser calculates that a 10,000-megawatt (10 GW) satellite power station, large enough to meet New York City's present power needs, would require a solar collector panel five miles square.

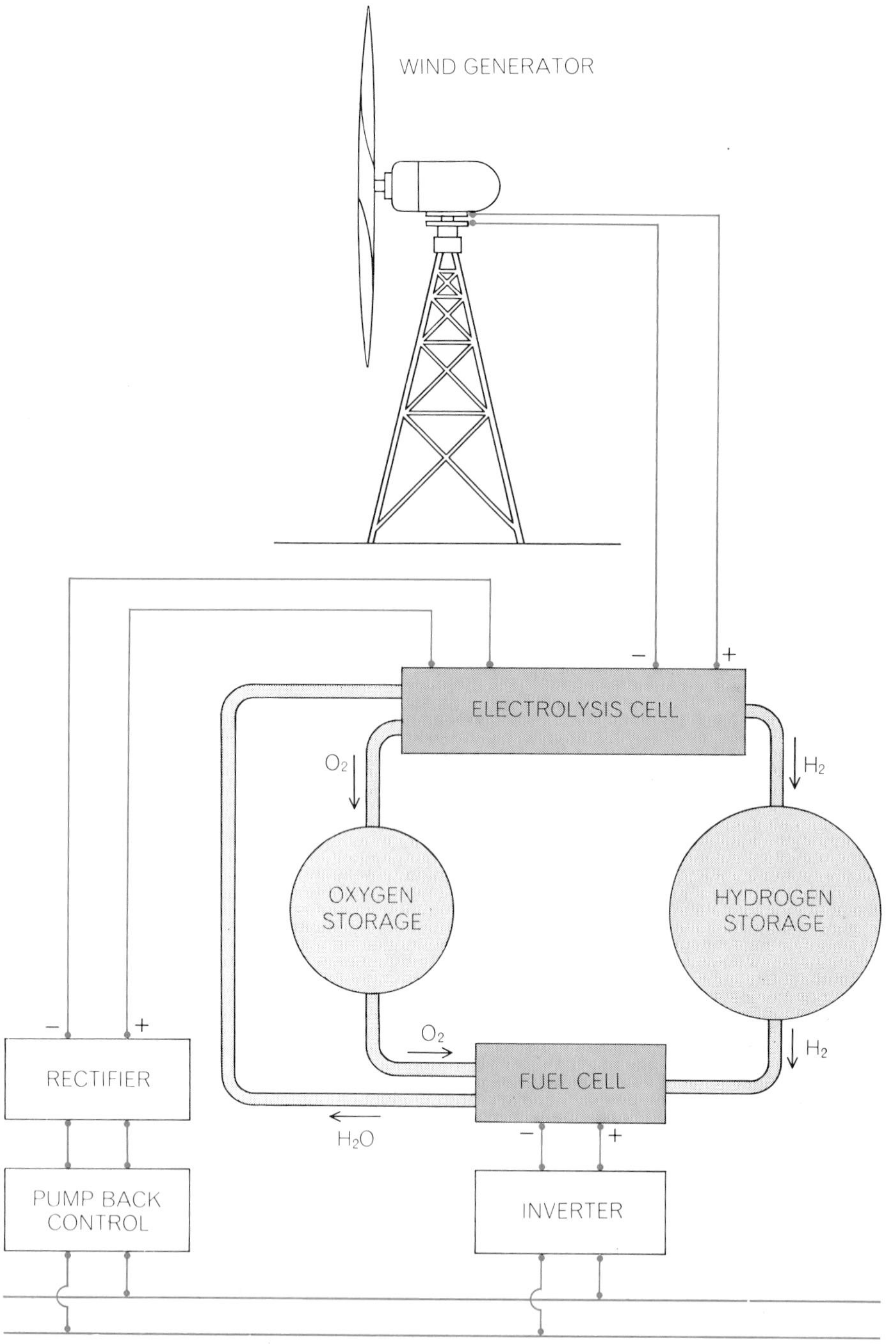

WIND GENERATOR
ELECTROLYSIS CELL
O_2
H_2
OXYGEN STORAGE
HYDROGEN STORAGE
O_2
H_2
RECTIFIER
FUEL CELL
H_2O
PUMP BACK CONTROL
INVERTER

WIND AS POWER SOURCE is attractive because it does not impose an extra heat burden on the environment, as is the case with energy extracted from fossil and nuclear fuels. Unlike hydropower and tidal power, which also represent the entrapment of solar energy, the wind is available everywhere. Unfortunately it is also capricious. To harness it effectively one must be able to store the energy captured when the wind blows and release it more or less continuously. One scheme would be to use the electricity generated by the wind to decompose water electrolytically. The stored hydrogen and oxygen could then be fed at a constant rate into a fuel cell, which would produce direct current. This would be converted into alternating current and fed into a power line. Off-peak power generated elsewhere could also be used to run the electrolysis cell whenever the wind was deficient.

The receiving antenna on the earth would have to be only slightly larger: six miles square. Since the microwave energy in the beam would be comparable to the intensity of sunlight, it would present no hazard. The system, according to Glaser, would cost about $500 per kilowatt, roughly twice the cost of a nuclear power plant, assuming that space shuttles were available for the construction of the satellite. The entire space station would weigh five million pounds, or slightly less than the Saturn moon rocket at launching.

To meet the total U.S. electric-power demand of 2,500 GW projected for the year 2000 would require 250 satellite stations of this size. Since the demand to 1990 will surely be met in other ways, however, one should perhaps think only of meeting the incremental demand for the decade 1990–2000. This could be done with about 125 power stations of the Glaser type.

The great virtue in power schemes based on using the wind or solar energy collected at the earth's surface, farfetched as they may sound today, is that they would add no heat load to the earth's biosphere; they can be called invariant energy systems. Solar energy collected in orbit would not strictly qualify as an invariant system, since much of the radiant energy intercepted at an altitude of 22,300 miles is radiation that otherwise would miss the earth. Only the fraction collected when the solar panels were in a line between the sun and the earth's disk would not add to the earth's heat load. On the other hand, solar collectors in space would put a much smaller waste-heat load on the environment than fossil-fuel or nuclear plants. Of the total energy in the microwave beam aimed at the earth all but 20 percent or less would be converted to usable electric power. When the electricity was consumed, of course, it would end up as heat.

To appreciate the long-term importance of developing invariant energy systems one must appreciate what exponential growth of any quantity implies. The doubling process is an awesome phenomenon. In any one doubling period the growth quantity—be it energy use, population or the amount of land covered by highways—increases by an amount equal to its growth during its entire past history. For example, during the next doubling period as much fossil fuel will be extracted from the earth as the total amount that has been extracted to date. During the next 10 years the U.S. will generate as much electricity as

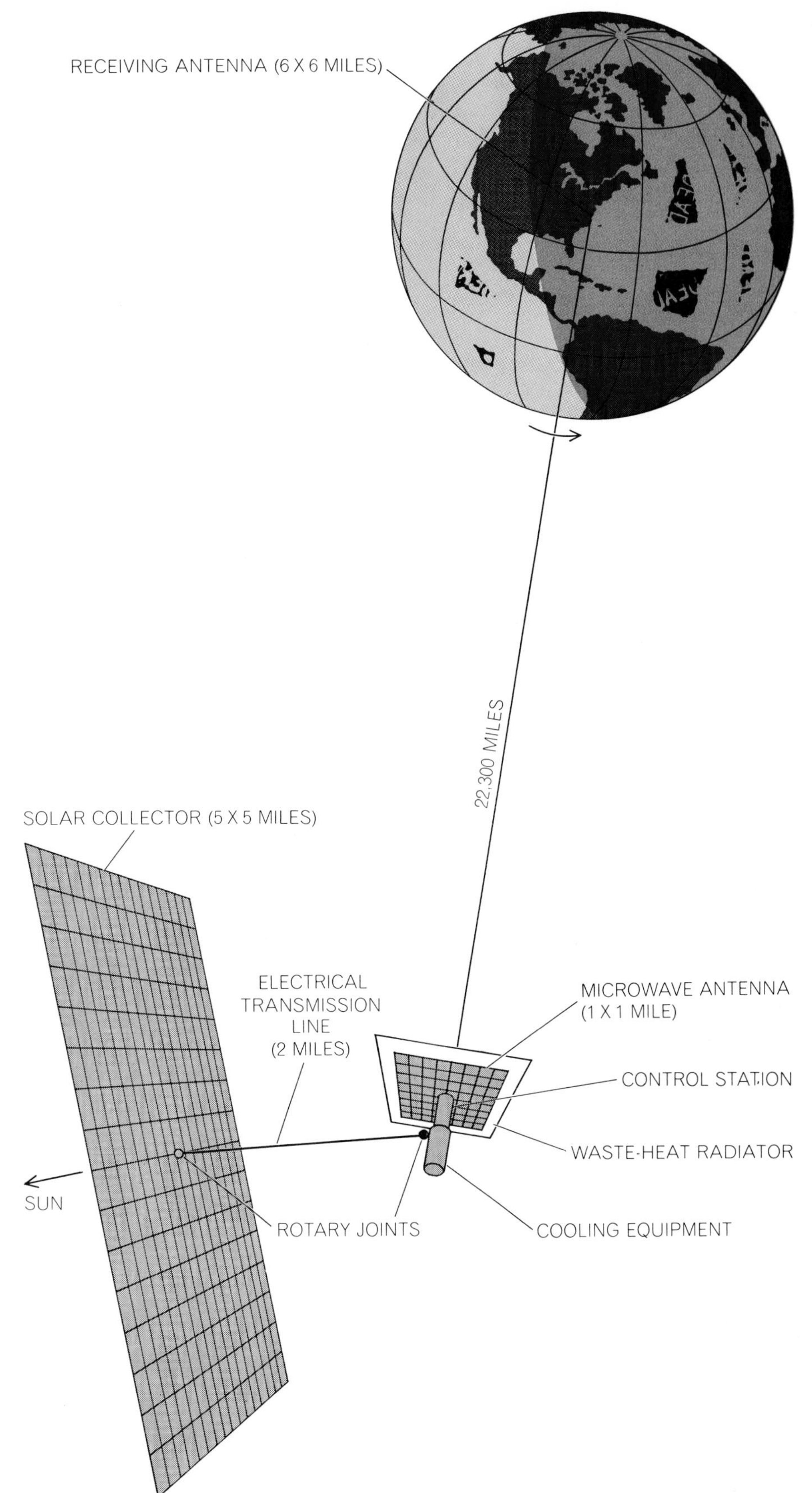

SOLAR COLLECTOR IN STATIONARY ORBIT has been proposed by Peter E. Glaser of Arthur D. Little, Inc. Located 22,300 miles above the Equator, the station would remain fixed with respect to a receiving station on the ground. A five-by-five-mile panel would intercept about 8.5×10^7 kilowatts of radiant solar power. Solar cells operating at an efficiency of about 18 percent would convert this into 1.5×10^7 kilowatts of electric power, which would be converted into microwave radiation and beamed to the earth. There it would be reconverted into 10^7 net kilowatts of electric power, or enough for New York City. The receiving antenna would cover about six times the area needed for a coal-burning power plant of the same capacity and about 20 times the area needed for a nuclear plant.

it has generated since the beginning of the electrical era.

Such exponential growth curves are usually plotted on a semilogarithmic scale in order to provide an adequate span. By selecting appropriate values for the two axes of a semilogarithmic plot one can also obtain a curve showing the number of doubling periods to reach saturation or depletion from any known or assumed percentage position [*see left half of middle illustration on page 102*]. As an example, let us assume that we have now extracted .1 percent of the earth's total reserves of fossil fuels and that the rate of extraction has been doubling every 10 years. If this rate continues, we shall have extracted all of these fuels in just under 10 doubling periods, or in 100 years. We have no certain knowledge, of course, what fraction of all fossil fuels has been extracted. To be conservative let us assume that we have extracted only .01 percent rather than .1 percent. The curve shows us that in this case we shall have extracted 100 percent in 13.3 doubling periods, or 133 years. In other words, if our estimate of the fuel extracted to this moment is in error by a factor of 10, 1,000 or even 100,000, the date of total exhaustion is not long deferred. Thus if we have now depleted the earth's total supply of fossil fuel by only a millionth of 1 percent (.000001 percent), all of it will be exhausted in only 266 years at a 10-year doubling rate [*see right half of middle illustration on page 102*]. I should point out that the actual extraction rate varies for the different fossil fuels; a 10-year doubling rate was chosen simply for the purpose of illustration.

In estimating how many doubling periods the nation can tolerate if the demand for electricity continues to double every 10 years (the actual doubling rate), the crucial factor is probably not the supply of fuels—which is essentially limitless if fusion proves practical—but the thermal impact on the environment of converting fuel to electricity and electricity ultimately to heat. For the sake of argument let us ignore the burden of waste heat produced by fossil-fuel or nuclear power plants and consider only the heat content of the electricity actually consumed. One can imagine that by the year 2000 most of the power will be generated in huge plants located several miles offshore so that waste heat can be dumped harmlessly (for a while at least) into the surrounding ocean.

In 1970 the U.S. consumed 1,550 billion kilowatt hours of electricity. If this were degraded into heat (which it was) and distributed evenly over the total land area of the U.S. (which it was not), the energy released per square foot would be .017 watt. At the present doubling rate electric-power consumption is being multiplied by a factor of 10 every 33 years. Ninety-nine years from now, after only 10 more doubling periods, the rate of heat release will be 17 watts per square foot, or only slightly less than the 18 or 19 watts per square foot that the U.S. receives from the sun, averaged around the clock. Long before that the present pattern of power consumption must change or we must develop the technology needed for invariant energy systems.

Let us examine the consequences of altering the pattern of energy growth in what may seem to be fairly drastic ways. Consider a growth curve in which each doubling period is successively lengthened by 20 percent [*see bottom illustration at right on page 102*]. On an exponential growth curve it takes 3.32 doubling periods, or 33 years, to increase energy consumption by a factor of 10. On the retarded curve it would take five doubling periods, or 50 years, to reach the same tenfold increase. In other words, the retardation amounts to only 17 years. The retardation achieved for a hundredfold increase in consumption amounts to only 79 years (that is, the difference between 145 years and 66 years).

Another approach might be to cut back sharply on present consumption, hold the lower value for some period with no growth and then let growth resume at the present rate. One can easily show that if consumption of power were immediately cut in half, held at that value for 10 years and then allowed to return to the present pattern, the time required to reach a hundredfold increase in consumption would be stretched by only 20 years: from 66 to 86 years [*see right half of top illustration on page 102*].

For long-term effectiveness something like a constant growth curve is required, that is, a curve in which the growth increases by the same amount for each of the original doubling periods. On such a curve nearly 1,000 years would be required for electric-power generation to reach the level of the radiant energy received from the sun instead of the 100 years predicted by a 10-year doubling rate. One can be reasonably confident that the present doubling rate cannot continue for another 100 years, unless invariant energy systems supply a large part of the demand, but what such systems will look like remains hidden in the future.

9

The Economic Geography of Energy

The Economic Geography of Energy

DANIEL B. LUTEN

The human uses of energy are reflected in patterns on the land. The prospecting, recovery, movement and ultimate use of energy resources are governed by the ratio of the benefit to the cost

All men have fire and have used it to change the green face of the earth, and those who live near fuel can have heat in abundance. Only those men who can convert heat and other forms of energy to work, and can apply that work where they will, can travel over the world and shape it to their ends. The crux of the matter is the generation of work—the conversion of energy and its delivery to the point of application. This article will explore some of the interrelations among the location of energy resources, the feasibility and cost of transporting energy commodities and the evolution of technology for converting energy.

Consider for a moment the three crucial developments of the past two centuries that have worked successive revolutions in the human utilization of energy. The first was the steam engine, invented and developed in England primarily as an answer to the flooding of deep coal mines by ground water. Removing the water was far beyond the capacity of human porters or of pumps driven by draft animals. For several centuries the task was accomplished by pumps driven by water mills. There was no realistic way, however, of conveying the action of water mills beyond the immediate site. Was coal mining to be forever confined to the streamside? The response to that challenge was the steam engine. It could operate wherever fuel could be delivered. In the 19th century its efficiency improved enough to make possible the steam locomotive, which could carry enough fuel with it to do work in transport.

The next big step came with the electric generator, the transmission of electricity and the electric motor, which freed work from its bondage to belts and shafts connected to the steam engine's flywheel; work could be provided wherever it was wanted, and in small or large amounts. The final step of this kind was the development of the automotive engine: a small power plant that was less convenient than an electric motor but was not even tied to a power line. Other fuels and conversion devices have appeared and will appear in the future, but they would seem to have less potential for working revolutions in our lives than the heat engine, small electric motors and the automobile.

Man's exploitation of an energy resource comprehends seven operations: discovery of the resource, harvest, transportation, storage, conversion, use and disposal. The discovery of the resource may be explicit and material, as in the case of a coal seam or an oil field. It may be conceptual: the idea of a reservoir or a scheme for capturing solar energy. Often it is the discovery of a conversion, as in the case of fire, the steam engine and uranium fission. And sometimes discovery comprises an entire series of technological improvements, as will be the case when shale oil is finally exploited successfully.

How much has resource discovery influenced human events? The U.S. ran on fuel wood until it had burned up the forests on croppable land as far as the prairies. England and Europe had done about the same thing, and when people ran out of wood, they turned to coal. (They complained; they preferred the old smells and smoke to the new, but they stayed warm with coal.) Whether it was the presence of coal that turned them to industry is another matter, one that is much more difficult to establish. Admittedly wood would not have sufficed, but a few lands with limited fuel have done well (notably Japan) and some with abundant fuel have not. Certainly local fuel does not seem to have been a sufficient condition, or even an entirely necessary one except for a pioneering society.

Gas and oil were adopted rather differently. It is said that as early as 1000 B.C. the Chinese drilled 3,000 feet down for natural gas, piping it in bamboo and burning it for light and heat and to evaporate brine for salt. Elsewhere candles persisted for millenniums and were only slowly succeeded by fatty oils in lamps. Coal had little to offer as an illuminant, but the coking of coal provided gas as well as coke. Handling gas required innovation, which came through the chemical studies of the late 18th century. In England "town gas" soon undercut the price of fatty oil for lamps in the new factories; in the less urban U.S. oil lamps persisted until kerosene appeared in the mid-19th century.

Discovery comes first in the exploita-

COAL FOR EXPORT passes through the yards of the Norfolk and Western Railway at Norfolk, Va. It is primarily high-grade metallurgical coal from fields in Virginia, West Virginia and Kentucky; Japan is the largest single customer. The yard can accommodate 11,520 coal hopper cars; a nearby yard handles another 9,880 cars. The two adjacent piers at lower left handle about 1,000 vessels a year. The pier at left, extending 1,870 feet into the Elizabeth River, is said to be the largest and fastest coal-loading facility in the world. It has two traveling loaders, each as high as a 17-story building, that can handle up to 8,000 tons of coal an hour. The system, consisting of car dumpers, conveyor belts and the loaders, combines coal of different kinds and grades, blending the shipments to order for the individual customers.

tion of a resource; use and then disposal are the next to last and last steps. The sequence of the intervening steps can vary depending on the resource and on the economics and geography and the specific set of operations they dictate. In some cases a preliminary conversion step is introduced: wood may be made into charcoal or coal into coke and petroleum must be refined.

A commodity can move by land either in a continuous process in a conduit or as a batch in a vehicle; shipping by sea must be by batches in vessels. The batch shipper has freedom of destination; a conduit constrains shipment to the chosen destination. The batch shipper, however, needs terminal storage facilities at both ends of every trip so that he can pick up and deliver his cargo

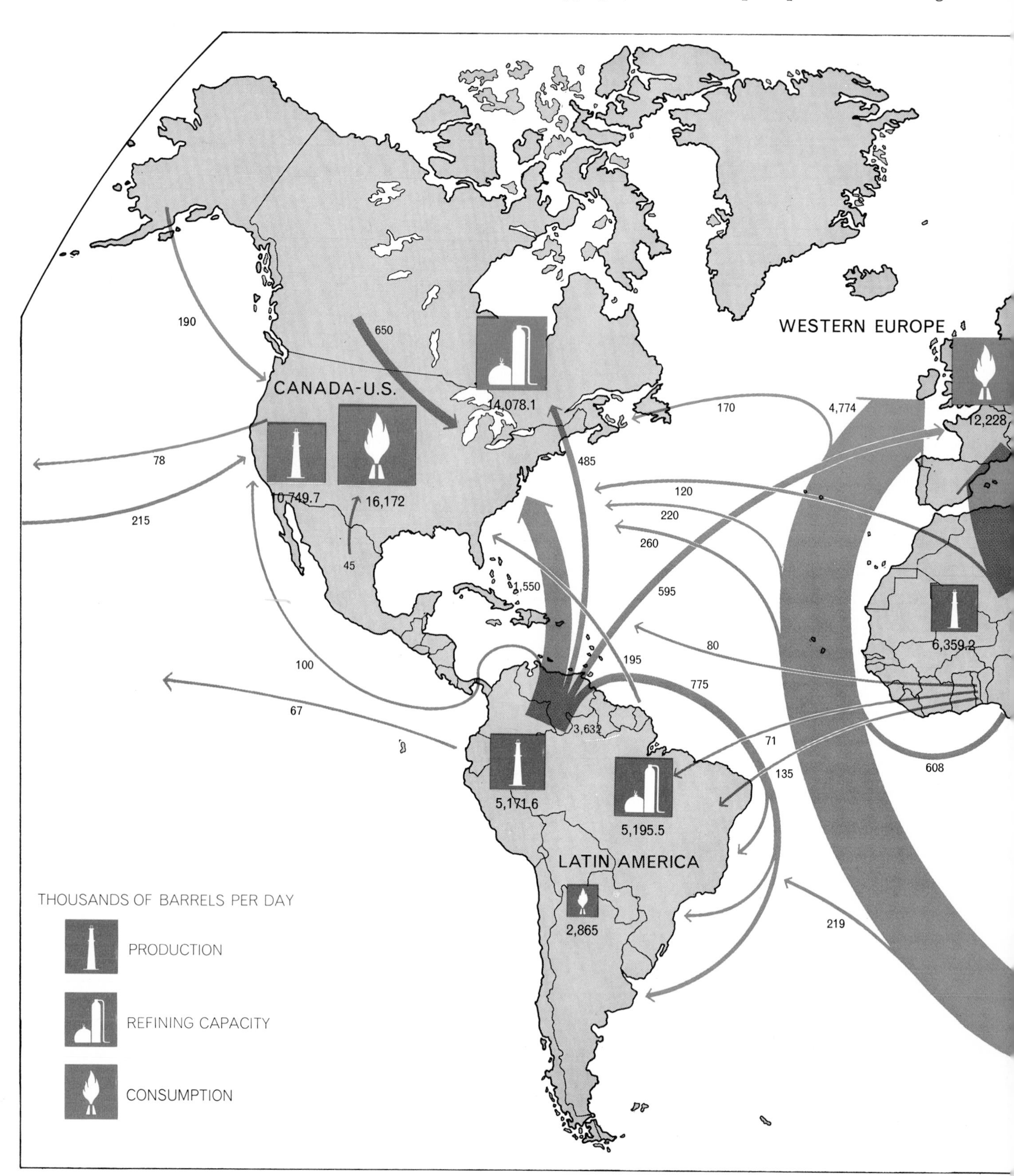

WORLDWIDE PATTERNS of oil production, refining, shipping and consumption are summarized by this map based on maps from the *International Petroleum Encyclopedia*. The data are for 1970. All quantities are in thousands of barrels per day. Export figures

with minimum lost time. For some commodities there are many possibilities; for others there are few choices or none.

The constraints on transport have had a significant effect on the adoption of new energy technologies. Primitive people can carry wood easily, coal less handily. The handling of liquids calls for pots and baskets; gases are uncooperative and elusive. Before the advent of simple and efficient equipment for containing and pumping fluids at high pressures, a development largely of recent decades, petroleum moved in barrels or in wood vats on flatcars, and long-distance transmission of gas was impractical. At sea, however, there were tankers, which began carrying oil from the Caucasus almost a century ago.

The combination of tankers and pipelines brought the fossil-fuel industry to

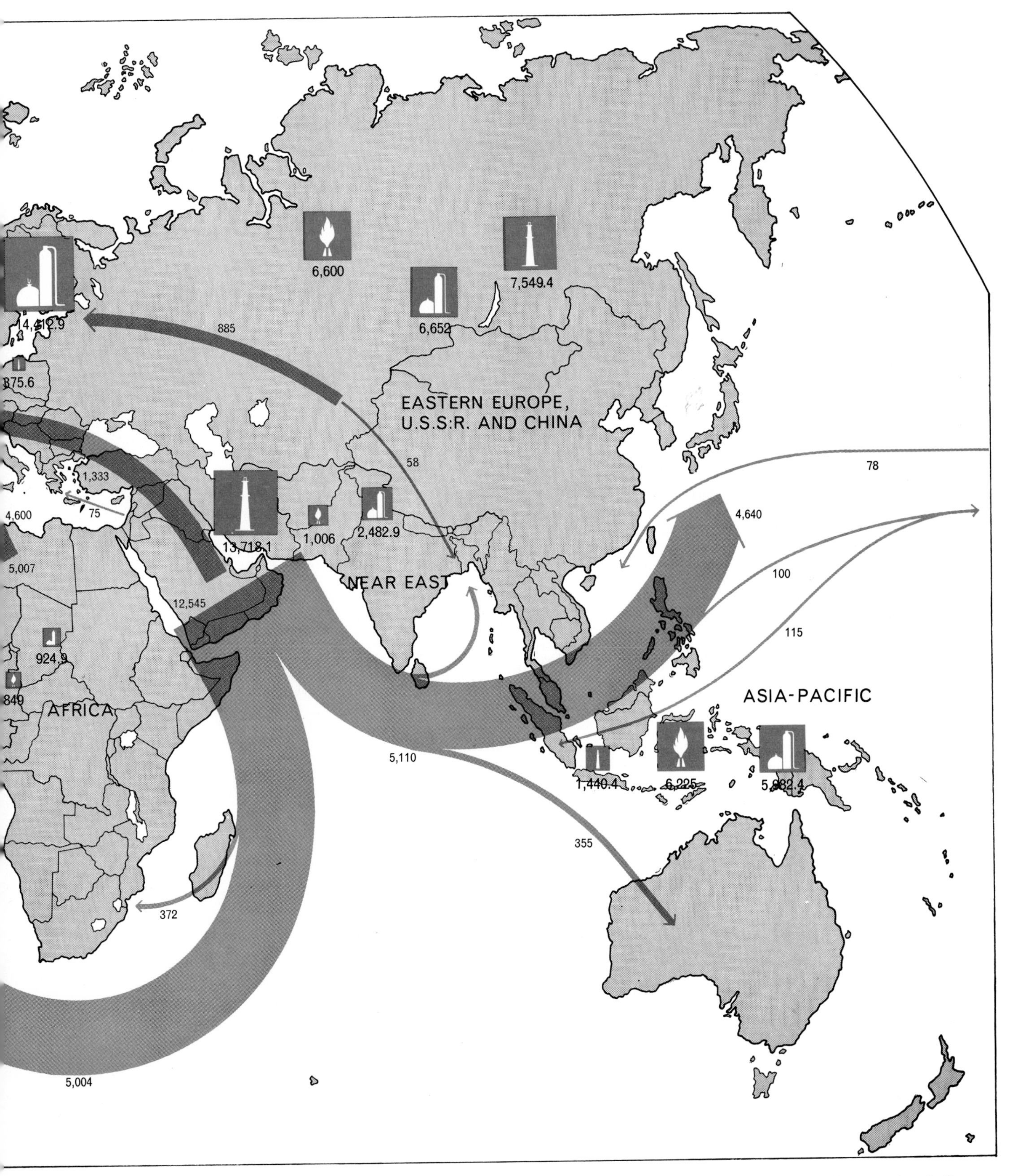

for eastern Europe, the U.S.S.R. and China refer only to exports from those areas to other parts of the world. The arrows indicate the origins and destinations of the principal international oil movements, not the specific routes. The U.S. is a heavy net importer.

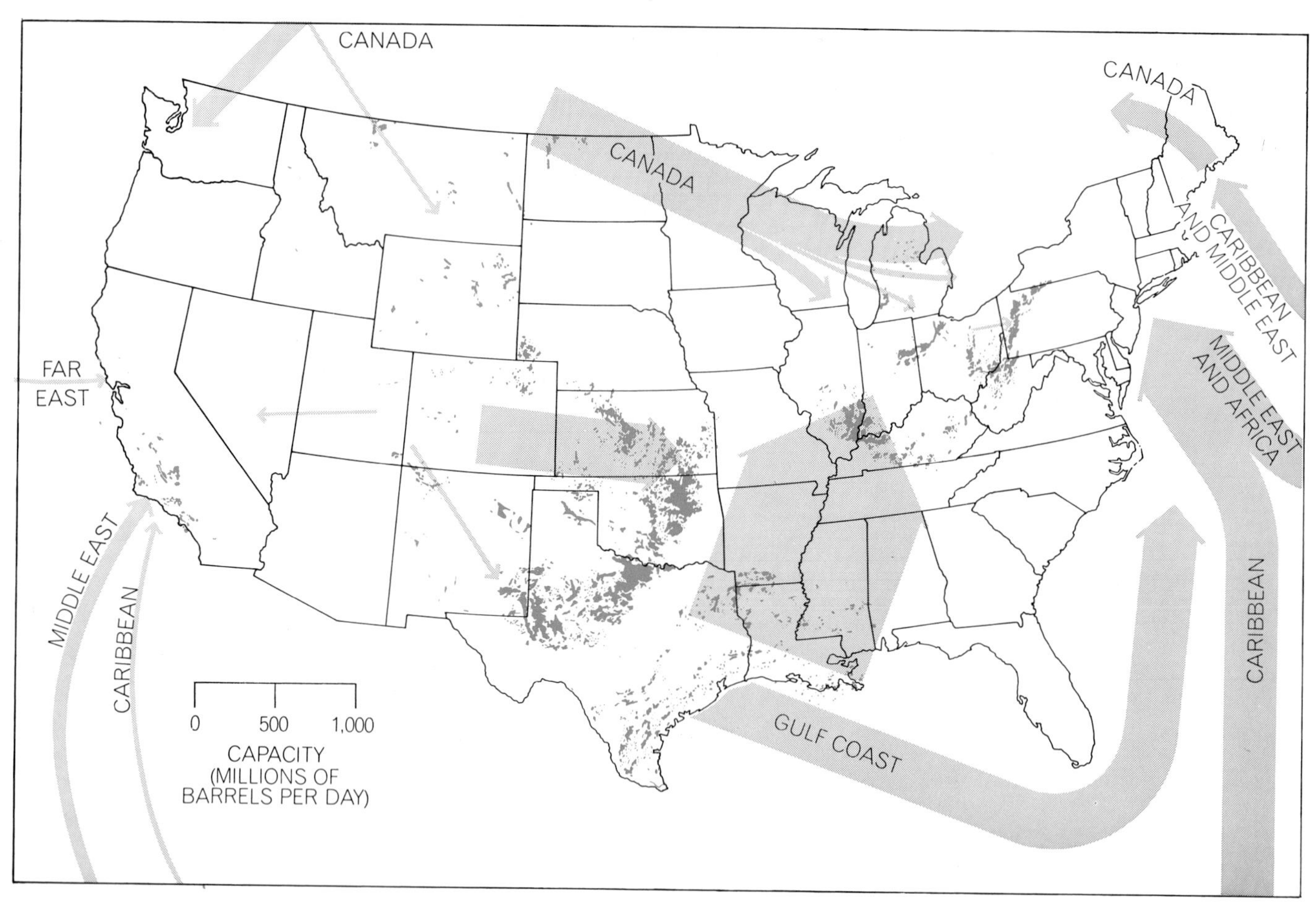

TRANSPORTATION OF CRUDE OIL to the U.S. and within the country is shown by a map adapted from the *National Atlas.* Data are for 1966. Arrow widths are proportional to movements by pipeline (*land*) and tanker (*sea*). Areas in solid color are oil fields.

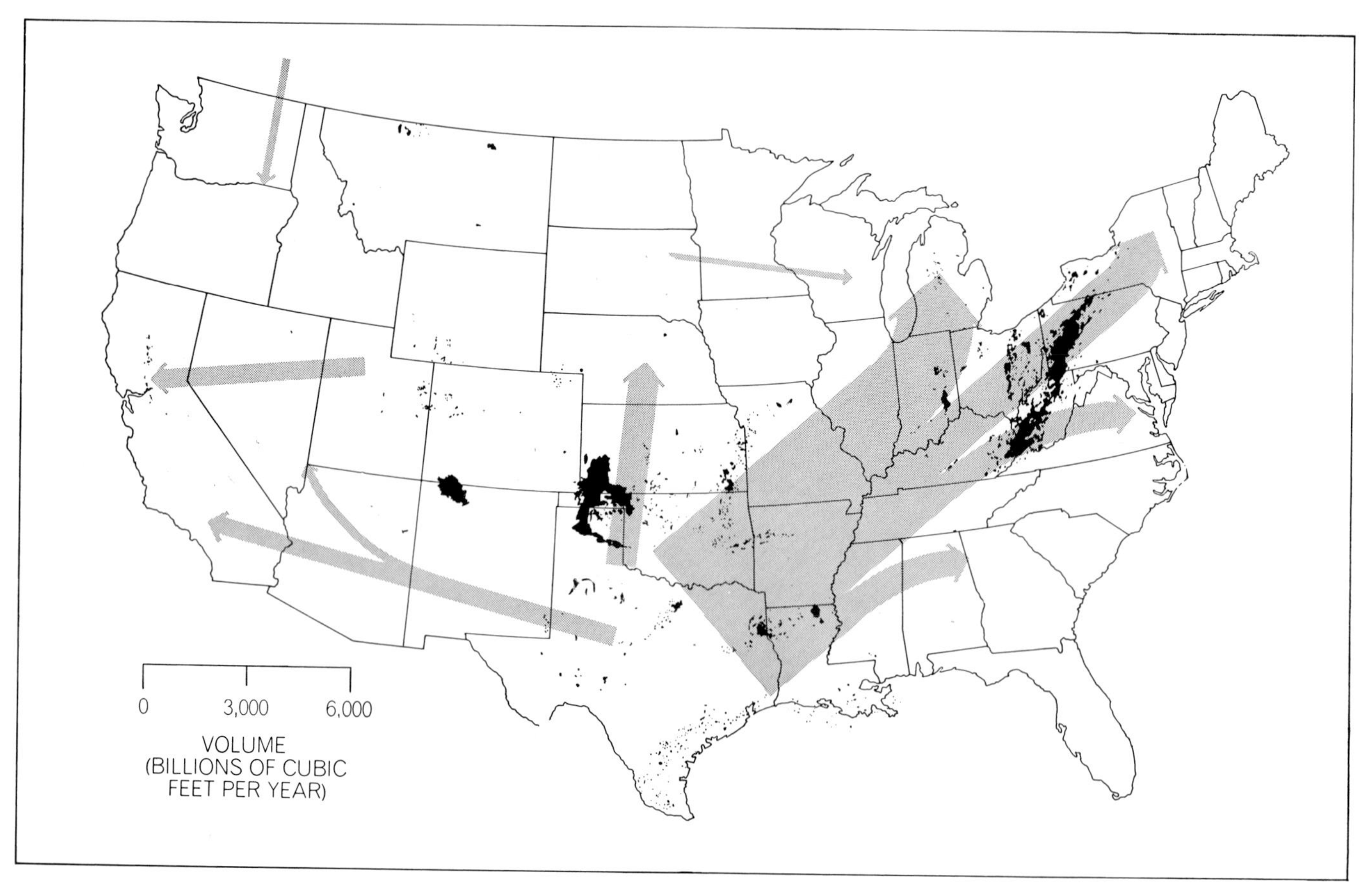

NATURAL-GAS MOVEMENTS are charted, based on figures for 1965. The development of techniques for transporting gas at high pressures in pipes has led to the sharp increase in the use of natural gas since World War II. Areas that are shown in black are gas fields.

a momentarily stable condition in the years after World War II. All the possibilities seemed to have been exploited. Now, with competition tightening, innovations are again being pressed hard and marginal improvements are being squeezed for any advantage. Oil brought great distances is made competitive by increasingly large tankers, but the million-ton supertankers now being proposed must be near the limit. Larger pipelines also shave costs, but most of the pipeline routes that have enough potential also raise international political issues; the proposed trans-Alaska pipeline has become a domestic political issue.

As technology advances, the feasibility of transporting some commodities improves. The fact remains that most commodities that can be transported at an acceptable cost today could also be transported economically long ago, although admittedly the distances have grown a great deal in this century. Some movements are still impossible; we do not know how to move electricity by sea, for example [*see bottom illustration on next page*]. The only recent real innovations in transport (except for the appearance of nuclear sources, with their trivial costs of transportation) are the movement of natural gas by sea as a refrigerated liquid and the development of new technologies for electrical transmission.

The power provided to any electrical-conversion unit is the product of the drop in voltage within the unit and the flow of current; the loss of energy as heat in a transmission line is the product of the square of the current and the resistance of the wire. Lower currents, higher voltages and larger wires (less resistance) therefore reduce waste. There are limits to the size of a wire, and so improvements in transmission were achieved primarily by utilizing alternating current (which could be transformed easily) and stepping up the voltage. Transmission voltages have increased as demands have grown and as transmission technology (insulation, for example) has improved, but the gains have required successive doublings of voltage rather than incremental increases, and the end of the road seems to have been reached for alternating current at less than 500,000 volts.

These gains, combined with the high growth rate of the electric-power industry in the U.S. and with the large economies of scale in the construction of power plants, have changed the look of the land. The oldest power plants were small

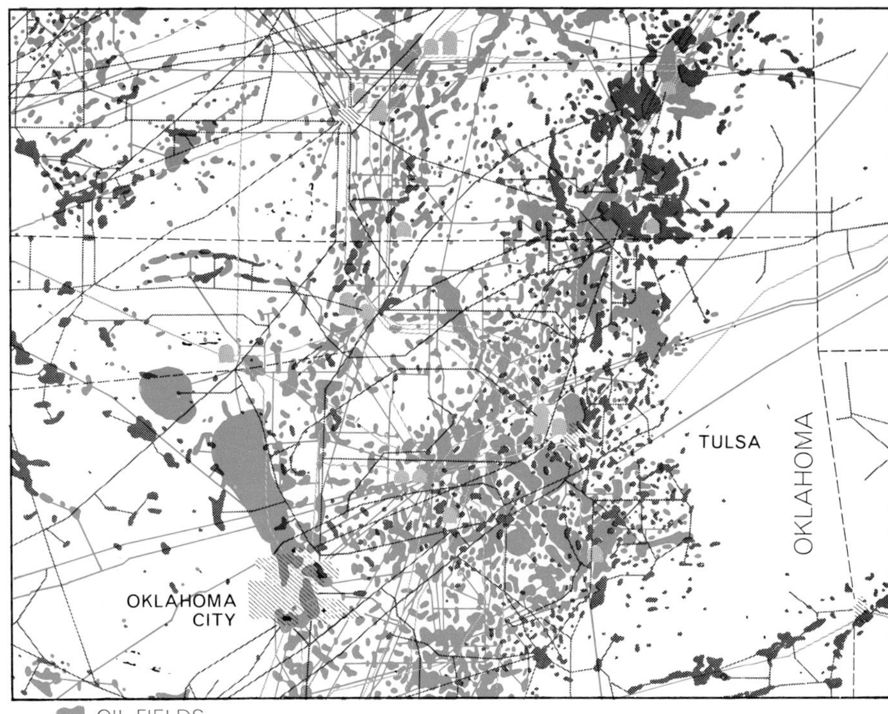

PIPELINES radiate from the rich oil fields and natural-gas fields of Oklahoma. Crude oil is piped from wells to refineries in the region or farther away; petroleum products from the refineries are piped to industrial and commercial centers, primarily in the Middle West.

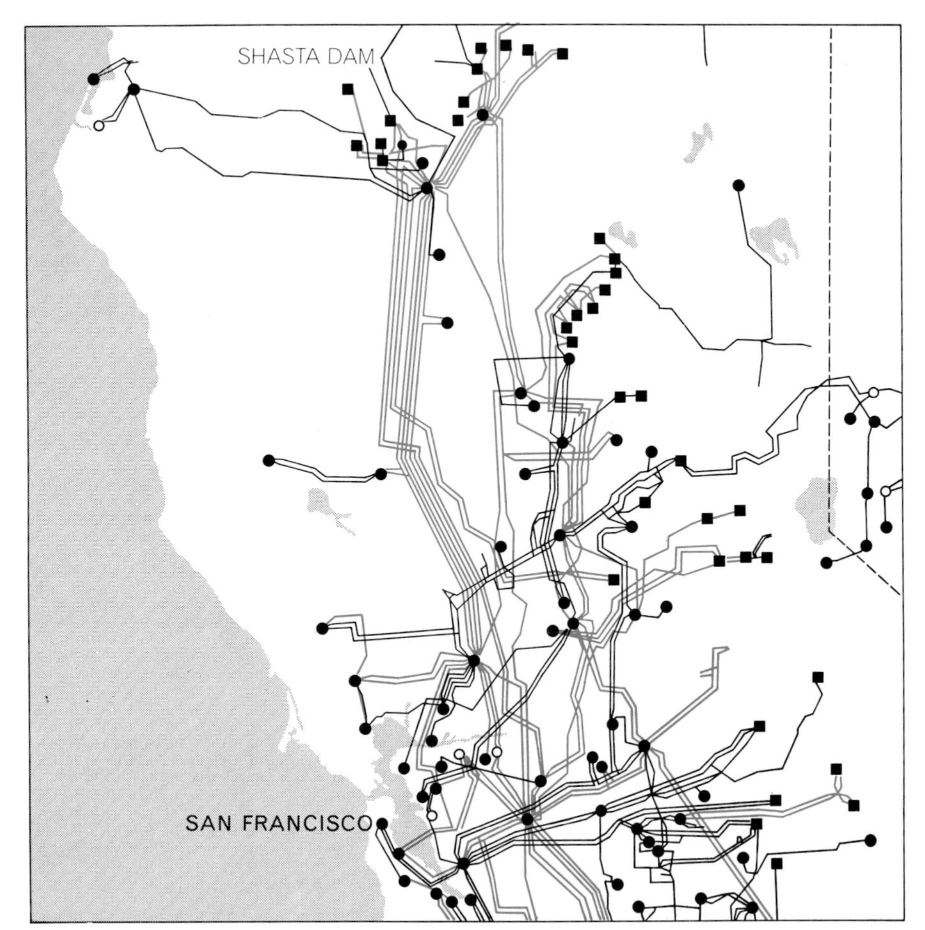

TRANSMISSION LINES radiate from power plants in northern California, the largest of which is Shasta Dam plant. Most of the electric power goes to the San Francisco area.

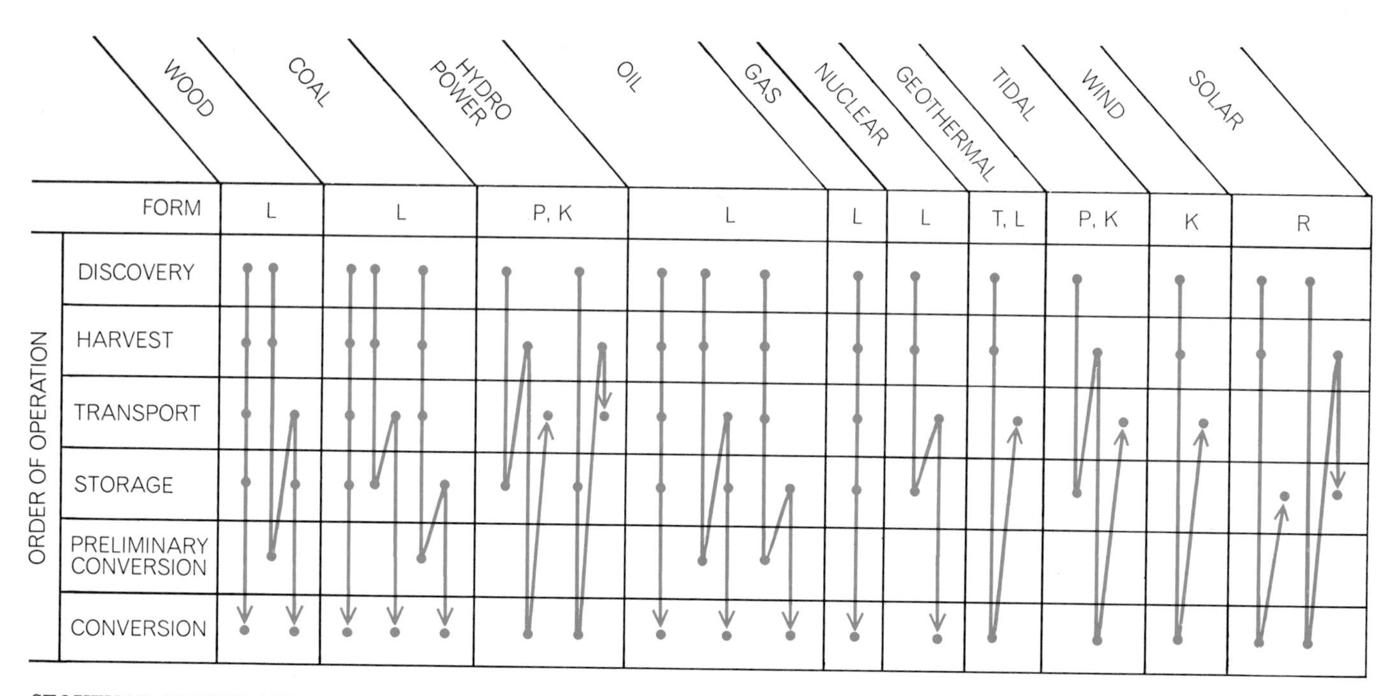

SEQUENCE OF OPERATIONS between the discovery and the use of an energy commodity is diagrammed. Energy is discovered in various forms: latent (*L*), potential (*P*), kinetic (*K*), thermal (*T*) or radiant (*R*). The resource is harvested, transported, stored and converted; sometimes there is a preliminary conversion step. The sequence of these steps varies for different commodities; in some cases there are alternate sequences. Arrows indicate the order in which the operations can be accomplished for 10 kinds of energy.

			WOOD	COAL	PETROLEUM	GAS	HEAT (STEAM)	ELECTRICITY	HYDRO POWER
LAND	BATCH	ARMLOAD							
		PACK							
		BASKET							
		POT							
		WAGON							
		TRUCK							
		RAIL		30	>15				
		VEHICLE FUEL TANK							
	CONTINUOUS	AQUEDUCT							
		PIPELINE			10	20			
		TRANSMISSION LINE						50	
		SLURRY PIPELINE		30					
SEA	BATCH	CARGO SHIP		<30					
		COLLIER		<30					
		BARGE		<30					
		TANKER			5				
		LNG TANKER				>20			
		SUPERTANKER			<5				

TRANSPORTATION of energy commodities can be by land or sea; on land it can be in batches or continuous, by sea only in batches. The colored boxes in the matrix indicate the feasible means of transport for each commodity. The numbers in some of the boxes give the approximate lowest cost for the major means of transport in cents per million B.T.U.'s for a 1,000-mile haul.

and widely scattered about the cities; the countryside had no electricity and no prospect of having it. Today power plants have become even larger; they are moving out of the cities, and high-voltage lines dominate miles of the countryside. Electricity is provided where it is wanted; transmission is not as cheap as moving fuel, and yet it is attractive to build big power plants and move electric power more than 100 miles to consumers. Still, the pressure for innovation continues. The privately owned public utilities that provide most of our electric power, even though they are entitled to prices that guarantee them a "fair profit" and are therefore in a sense free to rest on their laurels, are driven by their own imperatives to seek every possible increase in operating efficiency. (For one thing, as utilities lower their costs the public-utility commissions that set utility rates seem to lag in lowering the prices of electricity.)

The result is that even a trivial innovation may earn thousands of dollars a day, and the tendency is to judge its value by that potential rather than by its capacity to initiate a substantial revolution. Thus power companies adopt small improvements to alleviate some of the following inherent problems: (1) The demands of customers vary systematically by the time of day and the season, but unpredictable demands also arise and emergency shutdowns do occur. An isolated system must have enough spare equipment to handle such contingencies, but linking systems together with lines of high capacity makes some of the spare equipment unnecessary. (2) Peak demands are closely related to urban time schedules as well as to the sun. When a time-zone boundary is crossed, the period of peak demand shifts by an hour. Bringing in electricity from a neighboring time zone broadens the peak, reduces its magnitude and thereby again reduces the amount of generating equipment needed. (3) Some of the great hydropower facilities—Grand Coulee is the best example—can sell power very cheaply; others were built with the intention of selling power for premium prices, mostly at the hours of peak daily demand. Outlets for such peak-hour power may be many hundreds of miles away.

For all these reasons the power grids of the 48 states are now fairly well interlinked. It must be doubted that the resulting savings come to as much as 10 percent. Still, the interest in ever cheaper transport persists. Recently the devices of solid-state physics have provided means for transforming voltage (and current) with direct current. Because direct current is more tractable than alternating current at high voltages, utilities are now turning back from alternating to direct current and are beginning long-distance power transmission at extrahigh voltages (EHV) of 750,000. The next step may be the use of superconductors. All metals, when cooled to near the boiling point of helium, become superconductive, or quite without resistance to the flow of current. The use of superconductors could change the technical task involved in transmission dramatically, from the reduction of energy loss as heat to the operation of an elongated ultralow-temperature refrigerator. The first commercial application of superconductor transmission may be to bring power into urban areas too crowded for the wide corridors required for conventional high-voltage lines.

Storage presents its own set of constraints. Electricity is hardly storable as such in commercial quantities. Instead we resort to a subterfuge: building artificial reservoirs into which water can be pumped electrically and from which electricity can be retrieved by reversing the flow of water and letting the motors and pumps act as generators and turbines. Although this is quite efficient, it is a clumsy sort of thing; still, it is the best we can do. Storage batteries are not a substitute because they are expensive and have little capacity. One would think that about as much electricity could be stored in a battery as oil can be stored in a tank, because the same kinds of forces are being manipulated. Unfortunately reliable storage batteries are very heavy because they use chemical elements at the heavy end of the periodic table, notably lead, and provide only about as much energy as would result from an equal number of atoms at the light end of the series. (Clearly what is needed is a good storage battery in which lithium is oxidized and reduced instead of lead!) The same phenomenon gives electric automobiles an unsatisfactory performance and cruising range compared with automobiles that depend on hydrocarbon fuels.

The difficulties of storing electricity impose exacting constraints on the operation of electric-utility systems, as residents of many U.S. cities have learned in recent years. When customers demand more electricity by switching on lights or air conditioners or other machinery, the production of power must be increased to meet the demand. Little flexibility exists; electricity does not stretch or squeeze easily. To keep a system in balance requires minute-by-minute attention; at least, since electricity moves at a notably high velocity, increased production does reach the customer without delay.

If gas customers ask for a greater flow, on the other hand, gas will simply expand to a considerable degree within the pipeline and so meet the increased demand. Minute-by-minute flow is therefore no problem. The other side of the coin is that gas comes down the pipeline rather slowly, and so if the neighborhood supply runs short, it may take a day or two to make up the deficiency. Accordingly the marketers of gas usually have to arrange some kind of local storage. The large gasholders one sees on the outskirts of cities do not hold enough. To meet the possible peak demand for a day in the San Francisco Bay area, for example, would take a gasholder equivalent in volume to a cube 1,000 feet on a side; existing ones have perhaps 1 percent of that capacity. A common provision is therefore storage in depleted gas fields or in the transmitting pipeline itself. Gas is compressible, and if the upstream pressure is increased, not only can the gas be sent along faster but also larger amounts can be stored in the pipeline near the demand. A pipeline three feet in diameter running at 400 pounds per square inch contains about a million cubic feet of gas per mile, or a billion cubic feet per 1,000 miles—equal to the capacity of the 1,000-foot cube.

The third case is that of the supplier of liquid fuels. Here storage is so easy and so much of it is provided all along the distribution chain that no real technological problem remains. It is easier than keeping grocery shelves stocked.

In the synthesis of these unit operations both technological advances and economies of large-scale operation have contributed to lowering the cost of the alternatives for meeting demands. In general great economies of scale result only from the phenomena of liquid flow, which cause the capacity of a pipeline to increase as a high power of its diameter. One very different example of such economy is seen in strip-mining: the stripping away and movement of overburden is now being handled by outsized equipment, making operations economically attractive that would have been unacceptable a generation ago. Yet one suspects that here too, as in the case of supertankers, the end of the road of increasing scale is close at hand.

Energy is almost never harvested in

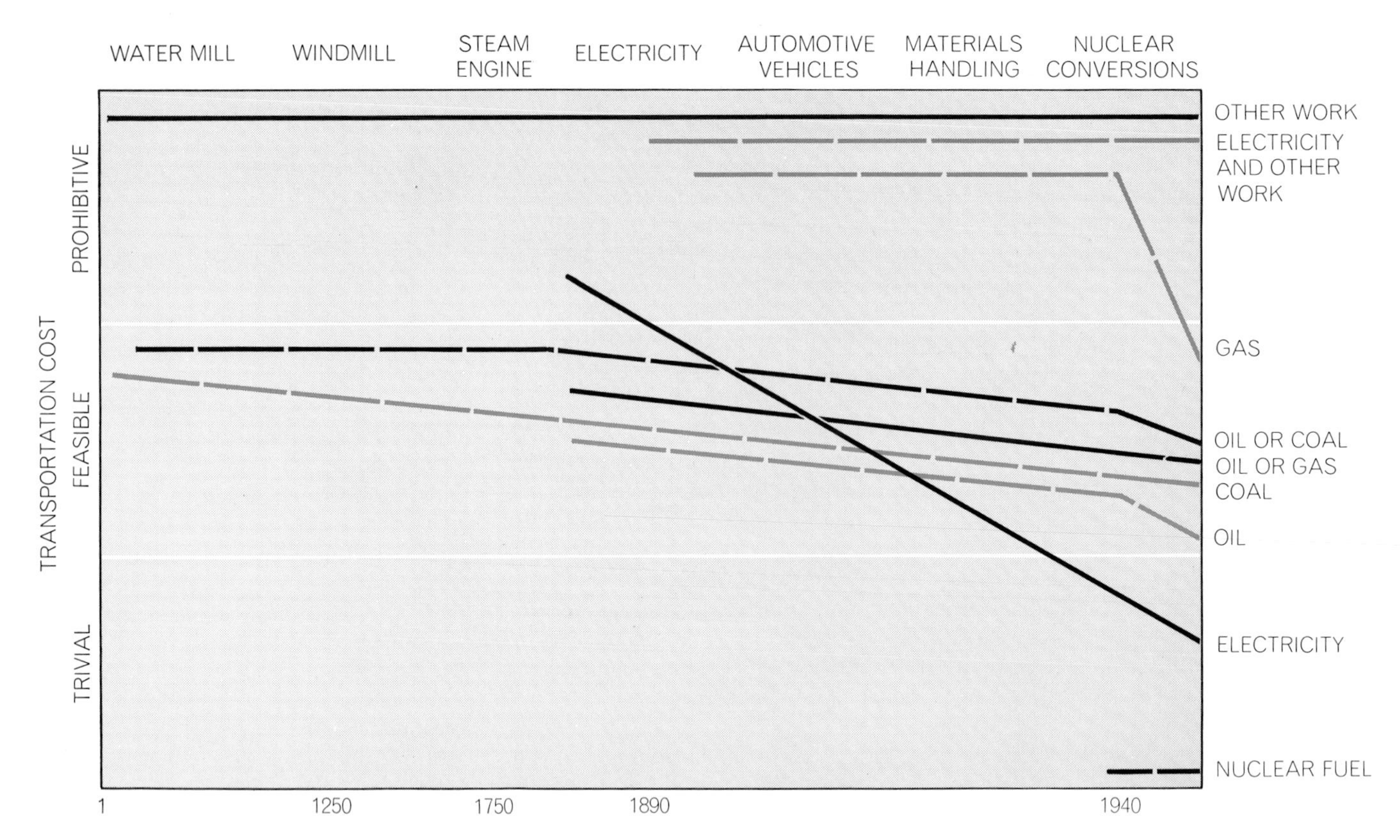

TRANSPORTATION COSTS may make it impossible to move some forms of work, such as wind or water power, from the site where they are developed. The costs of other commodities have varied through history; in many cases technological changes make a cost feasible that was once prohibitive. The curves relate the general level of costs for transportation by sea in batches (*broken colored lines*), by land in batches (*broken black lines*) and by continuous methods such as power lines or pipelines (*solid black lines*).

the form in which it is to be used, and therefore it must ordinarily go through a conversion step [see "The Conversion of Energy," by Claude M. Summers, page 95]. The most significant conversions are those of latent energy to heat through combustion, and of heat to electricity. Once energy is in the form of electricity all the gates are open, even though the toll through some gates is excessive.

Centuries of development, innovation and growth have built up an intricate pattern of physical facilities and economic relations that connect discovered and harvested resources with sites of conversion, use and disposal [*see illustrations on pages 110 through 111*]. For the most part the patterns reflect the movement of energy from resource sites to the homes and places of work of growing populations at the various times and in the various amounts and forms that are needed.

In a sense the customer has been king; he has received what he wanted when and where he wanted it. It might be argued, as a matter of fact, that societies in which energy costs have been excessive have simply not prospered. In the U.S. the consumer has usually paid the price and paid little attention to paying; in return the energy industry has ordinarily met his demands while asking for a very minor share of his income. To estimate that share is difficult because so much of it is paid indirectly and because one scarcely knows whether to apply retail or wholesale prices, what to do about gasoline taxes and so on. Very roughly, every American consumes each day about 15 pounds of coal (200,000 B.t.u.) for 10 cents; two and half gallons of petroleum, half of it as gasoline (350,000 B.t.u.), for 50 cents; 300 cubic feet of

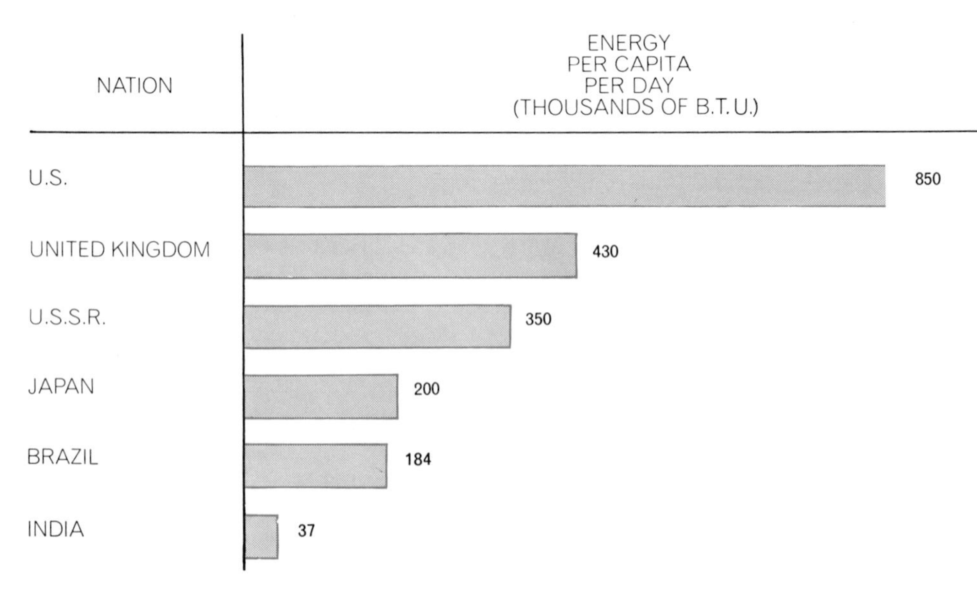

ENERGY PATTERNS are revealed by some international statistics. Energy per capita is about as expected, with a large advantage in the developed nations. (B.t.u. figures for Brazil and India would be about 150,000 and 22,000 respectively if "primitive" fuels were in-natural gas (300,000 B.t.u.) for 20 cents, and 24 kilowatt-hours of electricity for 45 cents. If one subtracts the 250,000 B.t.u. of the fuels that are used in generating the electricity, or 15 cents, the total comes to $1.10. Marked up to retail level, that would be about a tenth of the U.S. per capita personal income. Other inquiries have arrived at lower estimates, such as 4 percent or 7 percent, for the share of personal income spent on energy, but my own feeling is that a figure of 10 percent more nearly represents the situation from the consumer's point of view.

The resistance of the consumer varies. Two-thirds of his consuming is done for him in industry, commerce and transportation other than his own, and is beyond his direct control. It is hard to tell how much he resists buying industrial products, but his interest has been turning toward spending for services and it does seem that in some vague way his resistance to the purchase of industrial energy is increasing. In his home he behaves differently. No one can measure the extent to which he turns down the heat, turns off electric lights or skimps on gasoline, but the general impression is: not much. (Has anyone under the age of 30 ever turned off an electric light?) He pays a good deal more for electricity than he does for fuels but is easily persuaded to use electricity as a fuel, even though it costs him many times as much per unit of heat.

How about the rest of the world? First, it seems plausible, since fuels have long been available to men, that a highly technological society should show a high ratio of work to total energy, as expressed perhaps by kilowatt-hours per million B.t.u. Second, it can be argued that the construction of a thermal power system requires an intricate structure extending from mining through diverse forms of consumption, whereas the construction of a hydropower system (perhaps with assistance from a more technological society) can precede and is often intended to initiate development. Accordingly a high fraction of hydropower should be common in developing societies. Certainly the general experience is that the fraction of hydropower diminishes in the highly technological societies. Third, growth rates of electricity, for example, should be higher in the developing societies.

Such patterns do appear in the statistics but are far from infallible [*see illustration below*]. The U.S. is by no means the highest in kilowatt-hours per million B.t.u. In fact, it uses 35 percent of the world's electricity, just as it does with total energy. The reason is at least partly obvious: it is our excessively high consumption of gasoline for private automobiles. Brazil and India come in too high on the kilowatts-to-B.t.u. ratio, but the formalized statistics on which these numbers are based take no account of contributions from "primitive" sources: notably fuel wood in Brazil and cow-dung fuel in India. If these sources are counted in, the ratio drops from 32 for Brazil to six; for India it falls from 15 to six. (Perhaps as energy economies evolve the ratio should pass through a maximum and then decline.) The electrical growth rates are much what one would expect, except that Brazil's seem low. The hydropower percentages are generally in line, but they remind one not to forget climate and topography: Japan and the United Kingdom are both insular, mid-latitude and humid, but the former is mountainous, with a great many hydropower sites, and the latter is flat.

These, to be sure, are only the most superficial of the patterns associated with energy. Close examination of any society will reveal the influences on it of its particular experience with energy resources and energy conversion. The patterns one finds depend not only on such physical factors as the waxing and waning of resources but also on cultural variables: the development of technologies, changes in social patterns and the constraints of tradition, governmental policies and local fads and preferences. The resulting patterns are seldom simple, and it is particularly difficult to foresee the future. I should like merely to raise a few questions: Is the correlation between increasing use of energy and human welfare good enough, and is the hypothesis that more energy means a better life plausible enough, to warrant any hopeful extrapolation? Where on the rising consumption curve is the breaking point between gains and losses? Are we likely to find that point by encouraging growth until the customer—no longer interested in more energy or unable to afford it—finally offers resistance, and growth ends?

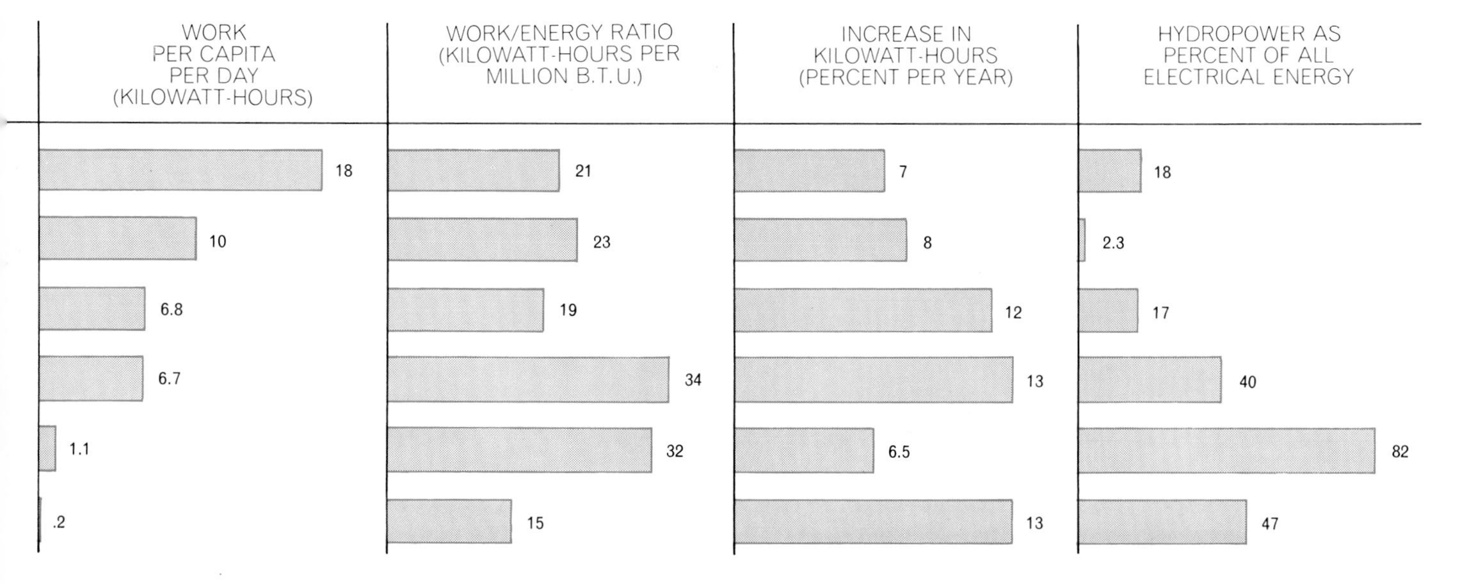

cluded.) One would expect kilowatt-hours per million B.t.u., a measure of the ratio of work to energy, to reflect technical expertise in the advanced countries, but the inefficiency of gasoline engines reduces the ratio there instead. The figures for hydropower's share of total electrical energy reflect not only the state of development (hydropower comes early) but also the geography of the countries.

10

Energy and Information

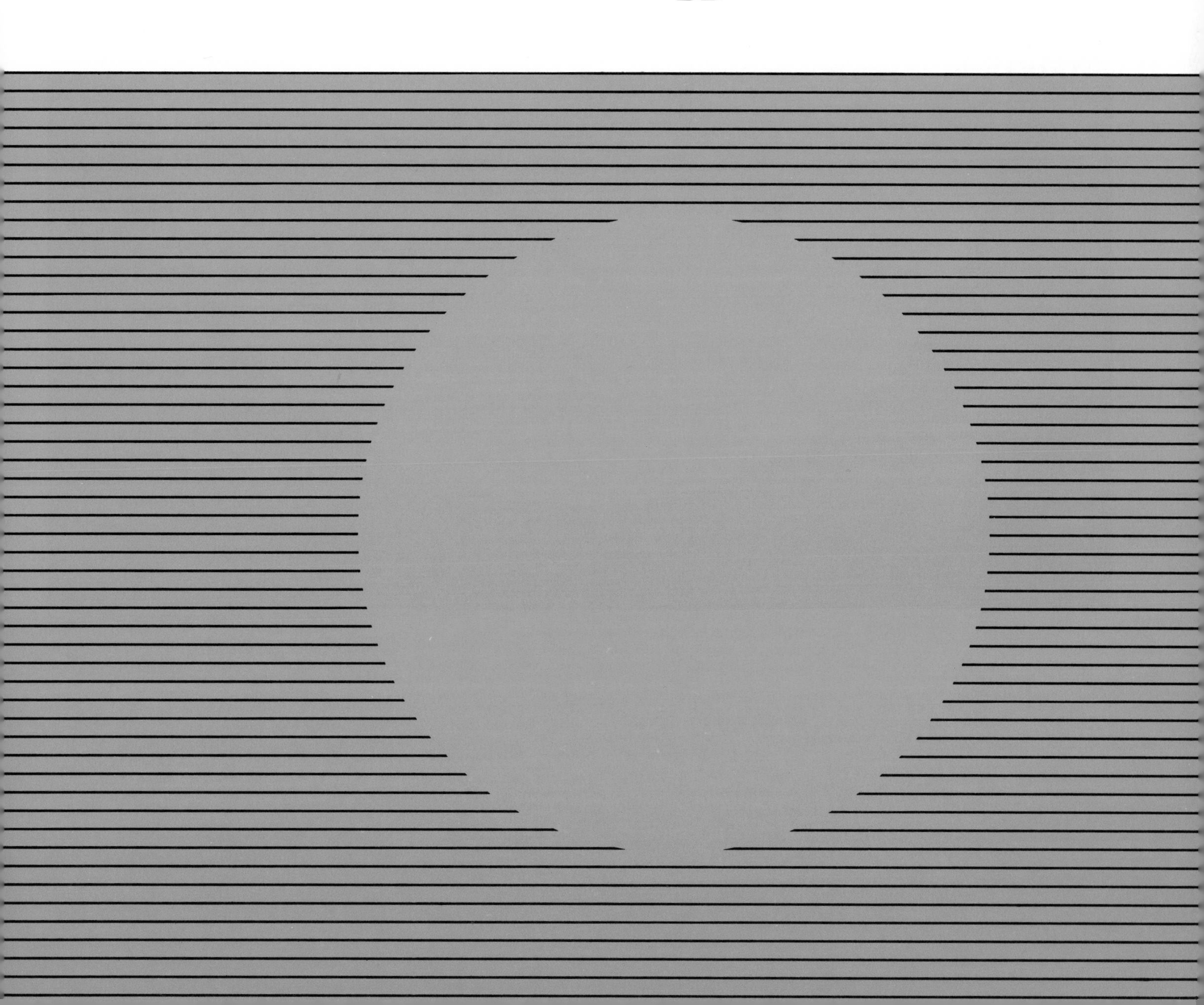

Photo by Robert R. Richardson

Energy and Information

MYRON TRIBUS AND EDWARD C. McIRVINE

The flow of energy in human societies is regulated by the tiny fraction of the energy that is used for the flow of information. Energy and information are also related at a much deeper level

Scientia potestas est—"Knowledge is power"—the Romans said, and 20 centuries later science has given an old phrase new meaning. The power to which the Romans referred was political, but that is a small detail. Science does not hesitate to give precise definition to everyday words such as "work," "power" and "information," and in the process to transform proverbial truths into scientific truths. Today we know that it takes energy to obtain knowledge and that it takes information to harness energy.

Research into the relation between energy and information goes back many years, but the era of precise yet general quantification of information began only with Claude E. Shannon's famous 1948 paper "The Mathematical Theory of Communication." It has long been known that any dynamic measuring instrument placed in a system must draw some power in order to actuate its mechanism. For example, a meter connected to an electrical circuit uses some power to cause the deflection of a pointer. The effect is reciprocal: whereas the theory of instrumentation shows why energy is needed to obtain information, recent advances in information theory show why information is needed for transformations of energy. In this article we shall pursue both lines of thought.

Ideas about energy are part of the education of every scientist. Since the concept of energy is discussed in detail elsewhere in this issue, we shall not repeat that discussion. The fundamentals of information theory are less well known. We shall therefore dwell at some length on the fundamental ideas of information before we take up the interaction of energy and information.

Ideas about probability play a central role in any theory of knowledge. In modern information theory probabilities are treated as a numerical encoding of a state of knowledge. One's knowledge about a particular question can be represented by the assignment of a certain probability (denoted p) to the various conceivable answers to the question. Complete knowledge about a question is the ability to assign a zero probability ($p = 0$) to all conceivable answers save one. A person who (correctly) assigns unit probability ($p = 1$) to a particular answer obviously has nothing left to learn about that question. By observing that knowledge can be thus encoded in a probability distribution (a set of probabilities assigned to the set of possibilities), we can define information as anything that causes an adjustment in a probability assignment. Numerous workers have demonstrated that Shannon's measure of uncertainty, which he called entropy, measures how much is expected to be learned about a question when all that is known is a set of probabilities.

Shannon's contribution to the theory of information was to show the existence of a measure of information that is independent of the means used to generate the information. The information content of a message is accordingly invariant to the form and does not depend on whether the message is sent by dots and dashes, by impressing a particular shape on a carrier wave or by some form of cryptography. Once this invariance is understood it becomes an engineering task to design a communications channel. One intriguing aspect of communication theory lies in the observation that the merit of a particular communications-channel design lies not in how well the actual message is sent but in how well the channel could have sent all the other messages it might have been asked to convey. A voltmeter that accurately indicates 1,000 volts when connected across a 1,000-volt potential drop is not of any value if it always reads 1,000 volts no matter what the actual potential is!

Although Shannon's measure of uncertainty was postulated for the purpose of designing better communications channels (and has served admirably for that purpose), it has much broader applicability. After all, any piece of physical instrumentation can be viewed as a communications system. Thus a probe (for example a thermocouple, a pressure transducer or an electrode) serves as a sender. Amplifiers, wires, dials and mechanisms serve as a communications channel. The human observer serves as a receiver. One can apply Shannon's ideas not only to the design of the apparatus but also to the code used for conveying the information. It is in the latter respect that the connection between information and energy is most interesting.

First we must define Shannon's measure. Suppose we have defined a question, denoted Q, and are uncertain of the answer. The statement "We have defined a question" needs to be made precise. We require that all possible answers be enumerated and that our confusion be over which of the possible answers is the correct one. If we ask something without knowing what the possible answers are, then we have not really posed a question; we have instead requested help in formulating a question. In order to define Shannon's measure we must deal with a well-defined question Q and have in mind a set of possible answers without necessarily knowing which answer is correct. (Suppose our question is: "Which number will turn up on this roulette wheel?" The possible answers consist of the numbers on the roulette wheel, and our uncertainty arises over which number to select.)

To make things compact we let the symbol X represent our knowledge about Q. (In our example X stands for all the things we know about the roulette wheel including our experience with the casino owner, the past history of the wheel and the actions of the disreputable-looking person standing near the table.) This knowledge, X, leads to an assignment of probabilities to the various possible answers. Assigning $p = 0$ to any one answer is the same as saying, "That answer is impossible." Assigning $p = 1$ to an answer is the same as saying, "That answer is certain." Unless X is of a very special nature we shall end up assigning intermediate values between 0 and 1 to all the possible results. Shannon's measure is represented symbolically by $S(Q \mid X)$ to emphasize that the uncertainty or entropy S depends on both the well-defined question Q and the knowledge X [*see upper illustration on this page*].

$$S(Q|X) = -K \sum p_i \ln p_i$$

ENTROPY IN COMMUNICATIONS was formulated mathematically by Claude E. Shannon in 1948. Shannon's entropy S is defined in terms of a well-defined question (Q) and knowledge (X) about Q. In Shannon's formula the symbol K represents an arbitrary scale factor, and the sign Σ means to "sum over," or simply add up, for each possible answer to the question Q the product of the probability (p_i) assigned to that answer and the "natural" logarithm of the probability ($\ln p_i$). Shannon went on to define the information (I) in a message as the difference between two entropies, or uncertainties: one that is associated with knowledge X before a message and the other that is associated with knowledge X' after a message; in symbols, $I = S(Q \mid X) - S(Q \mid X')$.

$$S' - S = \int_X^{X'} \frac{dQ_r}{T}$$

ENTROPY IN THERMODYNAMICS was defined by Rudolf Clausius in 1864 in terms of a "transformation" that always accompanies a conversion between thermal and mechanical energy. According to Clausius' formula, when a system changes from a state described by X to another state described by X', the entropy change ($S' - S$) is calculated by dividing each increment of reversible heat addition (dQ_r) by the absolute temperature (T) at which the heat addition occurs and adding the quotients over the change from state X to state X'; the integration sign ($\int$) symbolizes this mathematical operation. It can be shown that Shannon's function and Clausius' function are the same.

The mathematical definition of Shannon's entropy has the interesting property that if one correctly assigns $p = 1$ to one of the answers and (therefore) $p = 0$ to all the others, S is 0. (If you know the right answer, you have no uncertainty.) On the other hand, if all the probabilities are equal, S is a maximum. (If your information is so slight that you must assign equal probabilities, you are as uncertain as possible about the answer.)

In the preceding discussion we used the knowledge X about a question to define the entropy S regarding the uncertainty of the answer. Conversely, we could have used S to define X by saying that any X that maximizes $S(Q \mid X)$ is a state of maximum ignorance about Q. A man who does not know one answer from another is as ignorant about Q as he can possibly be. The only state of greater ignorance is not to know Q. Hence we can use the Shannon formalism to describe X quantitatively. Otherwise X is a qualitative concept.

For a given question (Q constant) it is of course possible to have different states of knowledge. Shannon defined the information in a message in the following way: A message produces a new X. A new X leads to a new assignment of probabilities and thus a new value of S. To obtain a measure of the information Shannon proposed that the information (I) be defined by the difference between the two uncertainties: in symbols, $I = S(Q \mid X) - S(Q \mid X')$.

The information content of a message, then, is a measure of the change in the observer's knowledge (from knowledge X before the message to knowledge X' after the message). A message that tells you what you already know produces no change either in knowledge (X remains the same) or in probability assignment and therefore conveys no information.

Shannon's measure is an invention. It was designed to fill a specific need: to provide a useful measure of what is transmitted on a communications channel. It has also been shown to be the only function that satisfies certain basic requirements of information theory. In the 23 years since Shannon put forward his measure thousands of papers have been written on the subject and no one has found a replacement function, or even a need for one. On the contrary, many alternative derivations have been found. We conclude that the Shannon entropy measure is fundamental in information science, just as the Pythagorean theorem is fundamental in geometry. Accordingly Shannon's concept of entropy should be a useful starting point for reasoning about information processes in general.

The word "entropy" had of course been used before Shannon. In 1864 Rudolf Clausius introduced the term in his book *Abhandlungen über die mechanische Wärmetheorie* to represent a "transformation" that always accompanies a conversion between thermal and mechanical energy. If a physical system changes from a state described by X (a particular combination of pressure, temperature, composition and magnetic field, for example) to another state defined by X' (a different combination of pressure, temperature, composition and magnetic field), then according to the Clausius definition, the entropy change is calculated by dividing each increment of heat addition by the absolute temperature at which the heat addition occurs and adding the quotients [*see lower illustration on this page*]. Except for the fact that Shannon's entropy and Clausius' entropy are represented by the same symbol and the same name, there appears at first sight nothing to indicate that the two functions are in fact the same function.

What's in a name? In the case of Shannon's measure the naming was not accidental. In 1961 one of us (Tribus) asked Shannon what he had thought about when he had finally confirmed his famous measure. Shannon replied: "My greatest concern was what to call it. I thought of calling it 'information,' but the word was overly used, so I decided to call it 'uncertainty.' When I discussed it with John von Neumann, he had a better idea. Von Neumann told me, 'You should call it entropy, for two reasons. In the first place your uncertainty function has been used in statistical mechanics under that name, so it already has a name. In the second place, and more important, no one knows what entropy really is, so in a debate you will always have the advantage.' "

The point behind von Neumann's jest is serious. Clausius' definition of entropy has very little direct physical appeal. It can be derived with satisfactory mathematical rigor and can be shown to have interesting and useful properties, particularly in engineering, but in a direct aesthetic sense it has not been satisfactory for generations of students. Simple physical arguments lead one to believe in the correctness of most quantities in physics. Surrounding Clausius' entropy there has always been an extra mystery.

The appearance of Shannon's measure, with the same name and the same functional representation as the earlier measure in statistical thermodynamics, aroused great interest among theoretical physicists. One of the best-known contributors to the subsequent discussion was Leon Brillouin, who treated the two entropies as the same in a series of papers and in the book *Science and Information Theory*. The proof that they are indeed the same (and not merely analogues) has been dealt with extensively elsewhere and will not be treated here.

The unit of information is determined by the choice of the arbitrary scale factor K in Shannon's entropy formula. If K is made equal to the ratio $1/ln\ 2$ (where the expression $ln\ 2$ represents the "natural" logarithm of 2), then S is said to be measured in "bits" of information. A common thermodynamic choice for K is kN, where N is the number of molecules in the system considered and k is 1.38×10^{-23} joule per degree Kelvin, a quantity known as Boltzmann's constant. With that choice for K the entropy of statistical mechanics is expressed in units of joules per degree.

The simplest thermodynamic system to which we can apply Shannon's equation is a single molecule that has an equal probability of being in either of two states, for example an elementary magnet. In this case both p_1 and p_2 equal 1/2, and hence S equals $+k\ ln\ 2$. The removal of that much uncertainty corresponds to one bit of information. Therefore a bit is equal to $k\ ln\ 2$, or approximately 10^{-23} joule per degree K. This is an important figure, the smallest thermodynamic entropy change that can be associated with a measurement yielding one bit of information.

In classical thermodynamics it has long been known that the entropy of mixing, per molecular weight of mixture, is a function of the fractional composition. Obviously the fractional concentration of a particular molecular species represents the probability of picking out a molecule of that species in a random sampling of the mixture. What does the act of mixing signify if we use the entropy of mixing as a measure of information? Imagine that we mix half a molecular weight of each of two isotopes. The resulting entropy change would be $N_0 k\ ln\ 2$, where N_0 (Avogadro's number, 6×10^{-23}) is the number of molecules per molecular weight. Numerically this change is about six joules per degree K., or 6×10^{23} bits. The latter number represents the number of

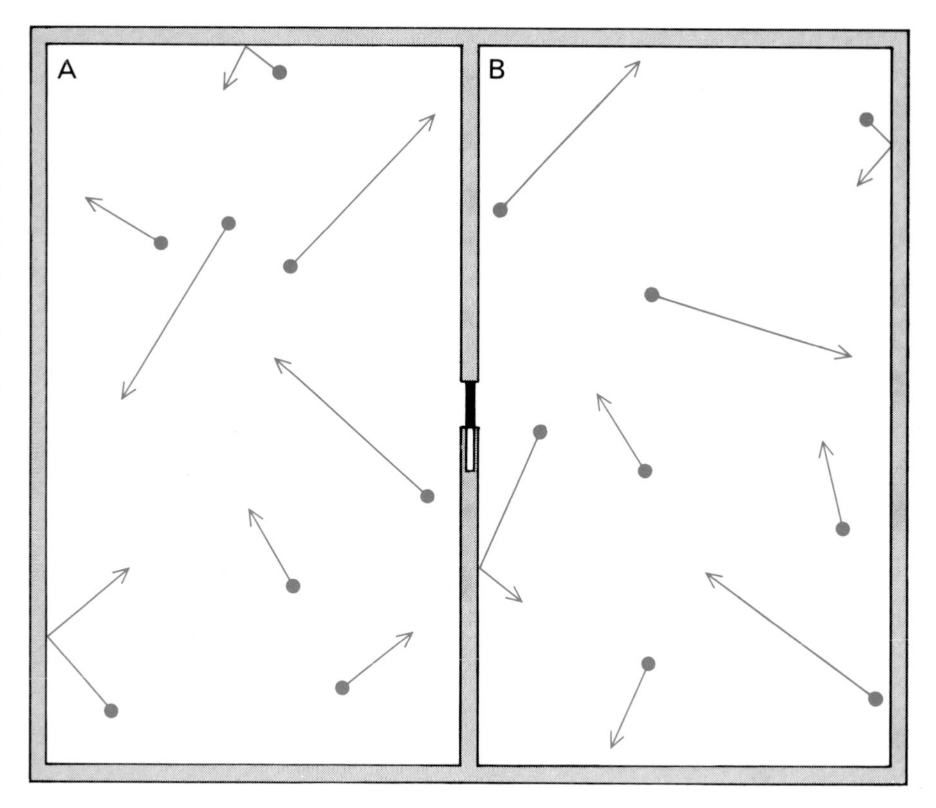

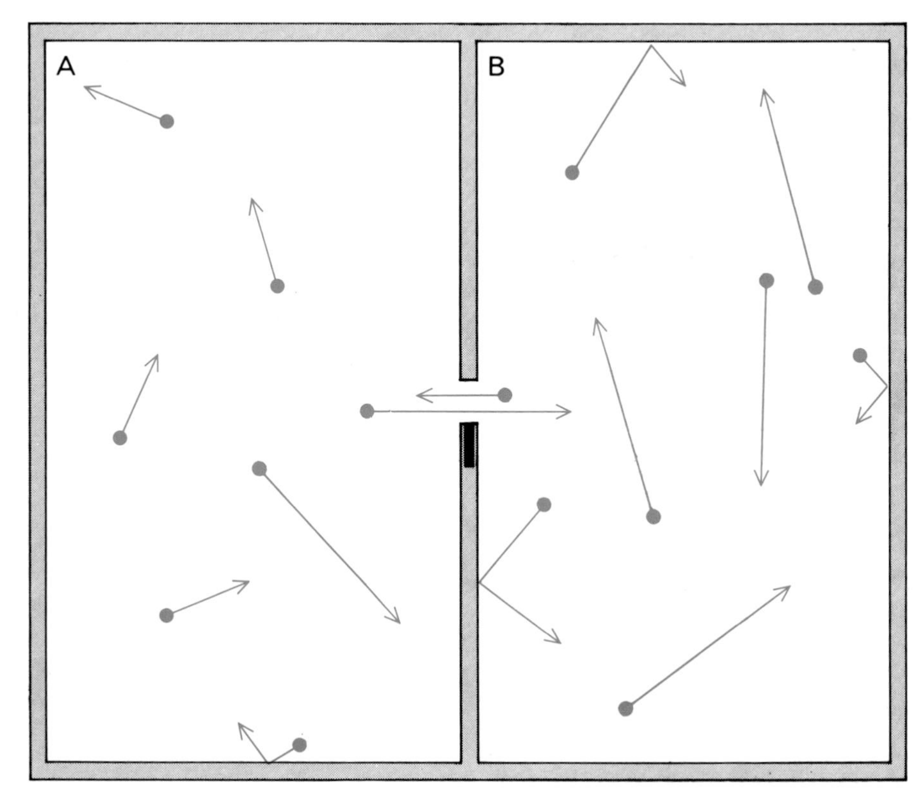

MAXWELL'S DEMON, a hypothetical being invoked by James Clerk Maxwell in 1871 as a possible violator of the second law of thermodynamics, was assumed to operate a small trapdoor separating two vessels full of air at a uniform temperature (*top*). By opening and closing the trapdoor so as to allow only the swifter molecules to pass from *A* to *B* and only the slower ones to pass from *B* to *A*, the demon could, without expenditure of work, raise the temperature of *B* and lower that of *A* (*bottom*), in contradiction to the second law of thermodynamics. The demon was finally "exorcised" in 1951 by Leon Brillouin, who pointed out that if the demon were to identify the molecules, he would have to illuminate them in some way, causing an increase in entropy that would more than compensate for any decrease in entropy such a being could effect. Without the input of energy represented by the illumination, the demon lacks sufficient information to harness the energy of the molecules.

decisions that would have to be made if a person were to sort the isotopes one at a time.

Just as the entropy of information has meaning only in relation to a well-defined question, so the entropy of thermodynamic analysis has meaning only in relation to a well-defined system. In our present understanding of physical science that system is defined by quantum theory. The question is: "In what quantum state is this system?" The answer is: "In some statistical combination of states defined by the quantum-mechanical solutions of the wave equation." In fact, these solutions define the possibilities we alluded to in discussing information and uncertainty. The probabilities encode our knowledge about the occupancy of the possible quantum states (possible, that is, for a given state of knowledge).

The concept of an inherent connection between the entropy of Clausius and the intuitive notion of information preceded Shannon's work by many years. In fact, the information-theory approach to thermodynamics is almost as old as thermodynamics itself. Clausius' 1864 book represents the earliest complete formulation of classical (nonstatistical) thermodynamics. By 1871 James Clerk Maxwell had introduced the role of information by proposing his famous demon [see "Maxwell's Demon," by W. Ehrenberg; SCIENTIFIC AMERICAN Offprint 317]. He suggested that a demon of minute size ought to be able to operate a small trapdoor separating two vessels, permitting fast molecules to move in one direction and slow ones in the other, thereby creating a difference in temperature and pressure between the two vessels [*see illustration on preceding page*]. Maxwell's demon became an intellectual thorn in the side of thermodynamicists for almost a century. The challenge to the second law of thermodynamics was this: Is the principle of the increase of entropy in all spontaneous processes invalid where intelligence intervenes?

From Maxwell's time on many leading investigators pondered the relation between observation and information on the one hand and the second law of thermodynamics on the other. For example, in 1911 J. D. Van der Waals speculated on the relation between entropy change and the process of reasoning from cause to effect. In 1929 Leo Szilard commented on the intimate connection between entropy change and information. In 1930 G. N. Lewis wrote: "Gain in entropy means loss of information; nothing more." Until Shannon came on the scene, however, there was no measure of information, so that the discussions could not be quantitative.

What Shannon added was the recognition that information itself could be given a numerical measure. If any of the early thermodynamicists had chosen to do so, he could have defined information to be consistent with thermodynamic entropy. After all, the "entropy of mixing" was well known. Any one of those men could have chosen to define information as the number of decisions required to "unsort" a mixture. The Shannon measure would have followed.

Shannon, who had no direct interest in thermodynamics, independently developed a measure of information. For practical reasons he chose to require the measure to meet certain logical criteria of consistency and additivity. In retrospect a logician can show that with these criteria Shannon was bound to produce a measure that would be consistent with thermodynamic entropy. Once it is recognized that the two subjects derive from common considerations it is straightforward to derive one from the other.

The Shannon formulation is somewhat more general since it is entirely a mathematical theory and is applicable

ACTIVITY	ENERGY (JOULES)	INFORMATION CONTENT (BITS)	ENERGY PER INFORMATION (JOULES PER BIT)
CHARACTER RECORD ACTIVITIES:			
TYPE ONE PAGE (ELECTRIC TYPEWRITER)	30,000	21,000	1.4
TELECOPY ONE PAGE (TELEPHONE FACSIMILE)	20,000	21,000	1
READ ONE PAGE (ENERGY OF ILLUMINATION)	5,400	21,000	.3
COPY ONE PAGE (XEROGRAPHIC COPY)	1,500	21,000	.07
DIGITAL RECORD ACTIVITIES:			
KEYPUNCH 40 HOLLERITH CARDS	120,000	22,400	5
TRANSMIT 3,000 CHARACTERS OF DATA	14,000	21,000	.7
READ ONE PAGE COMPUTER OUTPUT (ENERGY OF ILLUMINATION)	13,000	50,400	.3
SORT 3,000-ENTRY BINARY FILE (COMPUTER SYSTEM)	2,000	31,000	.06
PRINT ONE PAGE OF COMPUTER OUTPUT (60 LINES x 120 CHARACTERS)	1,500	50,400	.03

RATIOS OF ENERGY TO INFORMATION for various information-preparation, information-processing and information-distribution activities involving symbols are presented in this table. The energy values used are those typically involved in powering the mechanisms employed for these activities and in most cases are accurate only to about an order of magnitude. For character records information content is assumed to be about seven "bits" per character. The energy/information ratio for information systems based on character records and digitally encoded records varies from a few joules per bit down to a few hundredths of a joule per bit.

to all kinds of uncertainty. The thermodynamic theory is less general since it is bound to our "real world" environment of atoms, molecules and energy. Brillouin tried to emphasize this distinction by speaking of "free information" for the abstract Shannon quantity and "bound information" for the quantity when it described physically real situations (and thus thermodynamics). As we shall see in considering practical information-processing activities, the only real distinction is that natural physical situations involve much larger amounts of information than we appear able to control in our human-oriented information systems. There is no conflict between abstract Shannon information and thermodynamic information, as long as the questions we ask are physically real questions.

An interesting application of information theory to thermodynamics is provided by Brillouin's "exorcising" of Maxwell's demon. As we mentioned above, the demon led to apparent thermodynamic paradoxes. Brillouin pointed out that if the demon were to see the molecules, he would have to illuminate them in some special way. Since the black-body radiation in a gas vessel is the same in all directions, without a torch the demon would have no way of distinguishing the location of individual gas molecules. It takes special information to harness the energy of the molecules, and this information is above and beyond the normal thermodynamic information that serves to distinguish the system itself from its surroundings. Without the departure from equilibrium represented by the torch, Maxwell's demon lacks the information on which to act.

Unlike the demon, we do not live inside a gas vessel in equilibrium at a uniform temperature. Suppose for a moment that we did. Imagine that the earth is contained in a totally absorbing "black box" at a uniform temperature of 290 degrees K. (63 degrees Fahrenheit), a reasonable estimate of the real earth's average surface temperature. We would be as helpless as Maxwell's demon without a torch. In spite of the large energy flux and the moderate average surface temperature, an earth at equilibrium in an ambient environment would be inhospitable to life. No information could be processed; no energy would be available in the thermodynamic sense.

The actual case is of course different. The earth is part of a "sun-earth-space" system that is quite out of equilibrium. The sun plays the same role for us that the torch did for Maxwell's demon. By providing a departure from equilibrium it becomes a source of information and useful energy.

In considering the human use of energy and information, we must take into account the radiation balance of the earth's surface. The earth receives 1.6×10^{15} megawatt-hours of energy from the sun each year in the form of solar electromagnetic radiation, and it reradiates this energy principally as black-body radiation. Thus the earth approximately balances its energy budget. Man's use of energy on the earth's surface actually constitutes internal transactions with energy fluxes that are thermodynamically available, that is, usable before the energy is thermally degraded to the average surface temperature or chemically degraded by diffusion to the environment. Taking commonly accepted average values for the temperatures of the sun and the earth, the 1.6×10^{15} megawatt-hours of energy radiated to outer space carries with it the capability for an entropy decrease, or "negentropy flux," of 3.2×10^{22} joules per degree K. per year, or 10^{38} bits per second.

Of course, this negentropy flux derives from the energy flux from the sun to the earth to deep space. By storing the energy and negentropy in various systems (fossil fuels, lakes, clouds, green plants and so forth), the earth creates subsystems that are out of equilibrium with the general environment. In addition to the solar flux, then, we have the stored energy and negentropy of the earth's resources. Only in the case of the potential use of deuterium in fusion reactors does this stored energy exist in amounts significantly greater than one year's solar-energy flux. For practical purposes we may consider that both the energy flux and the negentropy flux at the earth's surface are due to solar processes. On the surface of the earth, therefore, the maximum steady-state rate at which information can be used to affect physical processes is of the order of 10^{38} bits per second. A great deal of this information is "used" in meteorological processes (cloud formation, thunderstorms, the establishment of high-altitude lakes and watersheds and so on). A large additional amount is "used" for the life processes of plants and animals. A comparatively small quantity is under the control of man, yet this quantity is responsible for man's technological reshaping of his environment.

The limitation of 10^{38} bits per second is not a stringent one. Consider the information rate of a television broadcast. Television stations broadcast 30 frames per second, each frame containing 525 scan lines. The resolution along each scan line allows about 630 bits of information to be encoded. The resulting information rate is $30 \times 525 \times 630$, or 10^7 bits per second.

Suppose we now consider a totally nonredundant television broadcast: one in which each dot is uncorrelated with the other dots on the same frame and each frame is unrelated to other frames. No human being could possibly absorb information from a television tube at such a rate. Even if the material being broadcast were a typical printed page (and on such a page there is a great deal of correlation between dots), it would take a person of reasonable skills about 60 seconds to absorb the information from one frame. Thus we can estimate the probable bit rate required to engage a human being in intellectual attention as being less than 10^4 bits per second. With a world population of less than five billion, the entire human race could have its information channels individually serviced and saturated with a bit rate of 5×10^{13}, very much less than the 10^{38} bits per second available.

Many old maxims point out that talk is cheaper than action. A comparison of the entropy balance and the energy balance on the surface of the earth indicates that the maxims are indeed a reflection of our experience. The amplification of information is easier than the amplification of power.

The total muscle-power output of the human race is estimated to be about 3×10^9 megawatt-hours per year (somewhat less than one megawatt-hour per person per year). Worldwide power usage under human direction is of the order of 7×10^{10} megawatt-hours per year, so that the energy-amplification ratio currently is about 25 to one. In the U.S. the amplification is much larger: approximately 250 to one. If all the thermodynamically available solar energy were used, an amplification of 500,000 to one is theoretically possible.

The possible amplification of information-processing activities is much greater. By eliminating the human operator from the chain of data-processing and machine control over physical systems, any direct dependence on the natural rate of human data-handling is removed. A human operator, working with a fixed set of questions, can use a modern digital computer to amplify his abilities by a factor well in excess of 10^6, perhaps by a factor of 10^{12}. The weak limitation on information-processing rates due to radiation to space (10^{38} bits per second) would indicate the theoretical maximum amplification to be in excess of 10^{24}.

It is worth observing that this great gap between the achieved and the achievable gives information technology a character different from that of materials technology or energy technology. In materials technology and energy technology scientists are accustomed to studying fundamental limitations and natural structures and engineers are accustomed to designing within a few orders of magnitude of these limitations. In information technology scientists find fundamental theorems not at all restrictive, and entrepreneurs discover that the freedom from constraints makes possible the construction of an almost totally new environment of information. Hence the advent of television programming, automatic telephone solicitation, computer-generated junk mail and other artifacts of an "information overload" culture. In the case of material and energy nature often cries "Halt!" to the changes wrought by technology. In the case of information man himself must issue the directives to ensure that technology is used for human betterment.

We have noted Brillouin's exorcising of Maxwell's demon. Beginning with this act Brillouin was led to investigate the relation between the entropy of an observation and the thermodynamic entropy, and he concluded that one bit of information requires $k \ln 2$ thermal entropy units. As Dennis Gabor once put it: "You cannot get something for nothing, not even an observation."

This comment has a special meaning for those of us who are engaged in the design of xerographic copying equipment. Among other reasons for concern about information and energy processes, we are interested in the minimum energy requirement for making a copy. The actual energy used at present is inconveniently high (mainly because the fixing of the final image takes about 90 percent of the power and involves thermally fusing the black toner to the paper), and so we shall not discuss the subject further. We discovered, however, that reading our copies often takes much more energy than making them. (A typical reading requirement might be 5,400 joules from a 90-second exposure of a 60-watt lamp.) Just as Maxwell's demon could not see molecules without a torch, so we cannot see images without illumination. Some luminous flux must be present involving electromagnetic radiation from a temperature greater than that of the environment. In the process of thermally degrading that energy a signal is generated. The distribution of reflectivity over the surface of the paper modulates the negentropy flux from the illuminator. Both the paper and the illuminator are required for the retrieval of information.

So far we have concentrated on the information needed to harness energy. Now it is time to examine the practical aspects of the energy requirements of information systems. The world of information technology includes both digital and analogue representations and the uses we make of them. The dramatic rise of the utility of digital computers sometimes leads us to overlook the other common representations of information: images and audio signals. For all three species of information representation we can consider the following distinguishable activities: the preparation of records, the storage of records, the processing of records and the distribution of records.

Information storage is a passive activity, and it does not intrinsically require the continuous input of energy (although in fact some forms of storage, such as a semiconducting digital-computer memory, do have power requirements). Information preparation, information processing and information distribution all require energy. The table on page 124 lists the information content and the energy in certain practical equipment configurations used for a number of activities involving character records and digitally encoded character records. Since the characters are chosen from a limited set, the information content is about seven bits per character. (In the case of nondigital characters we neglect

ACTIVITY	ENERGY (JOULES)	INFORMATION CONTENT (BITS)	ENERGY PER INFORMATION (JOULES PER BIT)
AUDIO RECORD ACTIVITIES:			
TELEPHONE CONVERSATION (ONE MINUTE)	2,400	288,000	.008
HIGH FIDELITY AUDIO RECORD PLAYBACK (ONE MINUTE)	3,000	2,400,000	.001
AM RADIO BROADCAST (ONE MINUTE)	600	1,200,000	.0005
PICTORIAL RECORD ACTIVITIES:			
TELECOPY ONE PAGE (TELEPHONE FACSIMILE)	20,000	576,000	.03
PROJECTION OF 35 MM SLIDE (ONE MINUTE)	30,000	2,000,000	.02
COPY ONE PAGE (XEROGRAPHIC COPY)	1,500	1,000,000	.002
PRINT ONE HIGH QUALITY OPAQUE PHOTOGRAPHIC PRINT (5″ x 7″)	10,000	50,000,000	.0002
PROJECT ONE TELEVISION FRAME (1/30 SECOND)	6	300,000	.00002

LOWER ENERGY/INFORMATION RATIOS generally exist for information activities involving audio and pictorial representations. This table overstates information content since the full bandwidth of available frequencies is never used in audio activities and the typical pictorial record contains a great deal of redundancy. As a result the energy/information values are in most cases accurate to less than an order of magnitude, but they are suggestive. It is clear, for example, that the information system that uses the smallest amount of energy per unit information is a hypothetical nonredundant television broadcast. Even this comparatively efficient information system uses for the purpose of communication only a tiny fraction of the thermodynamic information it requires to operate.

the possibility of added information in the form of changes of font, boldface, italics and other additions to sets of characters.) The energy is the energy involved in powering the mechanisms typically employed in these activities and in illuminating pages for reading. Most of the values are accurate to within about an order of magnitude. In the table we calculated the illumination from a 60-watt lamp. Obviously one can also read under a high-power arc lamp or by the light of a candle. The energy per information unit typically varies from a few joules per bit down to a few hundredths of a joule per bit.

Similar information is contained in the table on page 126 for practical equipment configurations used in activities involving audio representations and pictorial-image representations. We have used typical values for slide-projector illumination, radio and television receiver power and telephone central-office power. Clearly the values vary in individual instances. Additional uncertainties arise from our estimates of information content. Audio-intensity modulation, pictorial gray scale and color constitute multilevel coding techniques. For multilevel coding the information capacity of a channel is related to the signal-to-noise characteristics as well as to the frequency bandwidth. From a practical point of view, however, the table overstates information content; the full bandwidth is never used in audio activities and the typical pictorial record has a great deal of redundancy. The values are thus accurate to less than an order of magnitude, but they are nonetheless suggestive.

An extreme case of redundancy attends a 35-millimeter pictorial image of a page of print. Assuming a resolution of 100 line pairs per millimeter, about two million bits are used to represent approximately 3,000 characters. Most of the information is redundant and is used to convey the white spaces, the details of the character font and other material that may be of no importance to the message. Even a four-letter word could take up a million bits in a high-resolution photograph. At the other extreme a simple, nonredundant, two-level Baudot code can be used to represent the same four-letter word in only 24 bits.

Information technology has as one of its present concerns the best use of energy and physical structures to convey information as needed for human purposes, without undue redundancy and yet with veracity, style and taste. As a part of that concern the joint processes of energy flow and information flow are of special interest. We are led back to the consideration of the thermodynamic functions of a physical system that is involved in information processing. In the discussion that follows we shall make use of ideas recently developed by Robert B. Evans, now at the Georgia Institute of Technology, who has devoted a decade to the unraveling of the question.

The information-theory treatment of thermodynamics clarifies the concept of equilibrium. A few moments' thought should serve to convince one that the concepts "Distinguishable from the environment" and "Out of equilibrium" are the same. Our ability to recognize a system depends on the fact that it differs from its environment. "Thermodynamic information" is conceptually the same as "Degree of departure from equilibrium." If each of these quantities is measured in such a way as to satisfy the elementary properties of additivity, consistency and monotonic increase with the system's size, then apart from units of measure each will be the same mathematical expression, since they really refer to the same thing.

Thermodynamic information is defined as the difference between two entropies: $I = S_0 - S$. S refers to the entropy of a system of given energy, volume and composition. S_0 is the entropy of the same system of energy, volume and composition when it is diffused into (indistinguishable in) a referenced environment. It measures the loss of information in not being able to distinguish the system from its surroundings (as when an iceberg melts in the open sea).

The idea of using thermodynamic information as a generalized measure of the "availability" of energy was first put forward tentatively by Evans in 1965. (Although heat energy, mechanical energy and chemical energy can be converted into one another, they are not equally "available" to do work. What we call Carnot efficiency and Gibbs free energy were invented to deal with the availability of energy.) By 1969 Evans submitted his doctoral dissertation containing an entirely classical proof that a new quantity, obtained by multiplying his formula for thermodynamic information by an appropriate reference temperature, has most unusual properties. Evans has called this new function "essergy," for the essential aspect of energy [*see illustration at left*]. He has demonstrated that essergy is a unique measure of "potential work." Moreover, it incor-

$$S_0 = \frac{E + P_0 V - \sum \mu_{io} N_i}{T_0}$$

$$I = \frac{E + P_0 V - T_0 S - \sum \mu_{io} N_i}{T_0}$$

$$T_0 I = E + P_0 V - T_0 S - \sum \mu_{io} N_i$$

THREE EQUATIONS show the derivation of the concept of thermodynamic information. The top equation is based on classical thermodynamics: S_0 is the uncertainty when energy (E), volume (V) and the number of moles of various chemical species (N_i) are unrecognizable because they are distributed in an environment at a temperature T_0, a pressure P_0 and chemical potentials μ_{io}. The middle equation is derived from the top one on the basis of the relation $I = S_0 - S$, where I is information and S is the uncertainty about the system formed with energy E, volume V and composition N_i, and the system is now discernible from the environment. These equations were derived by Robert B. Evans, now at the Georgia Institute of Technology, in 1969. He showed that a new quantity obtained by multiplying the middle equation by T_0 is the most general measure of disequilibrium or "potential work." Evans has named this new quantity "essergy" (for the essential aspect of energy).

porates as special cases all previously known measures of departure from equilibrium (such as Gibbs free energy, Helmholtz free energy, the function used in Germany under the name "exergy," the Keenan availability function and so on). One can also think of thermodynamic information as a fundamental quantity representing the "signal-to-noise ratio" for a system in an environment T_0. Recall that kT_0 is the thermal-noise energy per degree of freedom, and that Evans essergy is T_0I. When I is of the order of $100kT_0$ or less, one is dealing with information processing. When I is larger, one is dealing with work and power. For example, in the flux of electromagnetic essergy the flux of essergy works out to be the same as what is called the Poynting vector. Close to a radar antenna the flux of essergy is large enough to cook a man; far away it becomes a weak signal.

Evans' essergy function has proved useful in analyzing power cycles, particularly for economic optimization. After all, energy can be neither created nor destroyed, so that an energy balance for a steam power plant is not very enlightening. An essergy balance, however, enables one to track down specific plant inefficiencies and see what they cost. Looking at the earth as a whole, that solar energy flux of 1.6×10^{15} megawatt-hours per year would be useless if its reradiation to outer space were not accompanied by an entropy flux of 3.2×10^{22} joules per degree K. (The corresponding change in essergy is -2.6×10^{15} megawatt-hours per year.) Ideally an essergy analysis could be performed on the natural processes of the earth's surface. This would form a foundation for ecologically sound planning of energy utilization by man, since it would provide an indication of the disturbance caused by proposed large-scale changes. The usefulness of the essergy function in practical matters has already been demonstrated in a comparative analysis of methods of making fresh water from seawater and a generalized study of power plants.

If we return to the examples of information-systems activities listed in the tables on pages 6 and 8, it is now possible to compare energy and information fluxes by comparing the contributions to essergy change. The example that used the smallest amount of energy per unit information was the nonredundant television broadcast, with 300,000 bits of information at a cost of six joules. Yet that information change represents a contribution of only 1.3×10^{-15} joule of essergy. Indeed, in terms of energy requirements the information-processing aspect of a television broadcast is a sidelight. What one is mostly doing with the energy is heating the room. A similar analysis would show that each example of what we have called an information-systems activity is in fact an energy activity that carries with it a small amount of information. Not until information technology reaches the state of handling information at one bit per molecule will we be addressing real information processes in the thermodynamic sense. Perhaps it is this wide gap between the amounts of entropy that are measured in practical information systems and the entropy of physical systems that has led to a reluctance on the part of classical thermodynamicists to adopt the information-theory view of the foundations of thermodynamics.

It has been a decade since the appearance of the first textbook to develop thermodynamics on the basis of information theory. The interval has not seen a conversion from the classical tradition. The classical treatment seems more entrenched than ever, to judge from recent conferences of teachers of thermodynamics. In view of the ever increasing importance of information technology in the scientific world, this is unfortunate. Perhaps the reason lies in the comments of the Russian worker A. I. Veinik, who wrote in 1961 apropos of his own attempt to restructure thermodynamics: "The traditional separation [into the] branches of physics is due to historical reasons rather than to the nature of the subject matter. A logical development of science must lead inevitably to the unification of these fields, and such a unification has not yet come about only because the century-old traditional outer trappings adorning the queen of the sciences, classical thermodynamics, have been defended with particular zeal. Apparently a large part of this defense has been based on the authority of the geniuses who brought this queen of sciences into existence."

Whatever the short-term outcome in educational circles, it is certain that the conceptual connection between information and the second law of thermodynamics is now firmly established. The use of "bound information" in the Brillouin sense of necessity involves energy. The use of energy, based on considerations of thermodynamic availability, of necessity involves information. Thus information and energy are inextricably interwoven. We may well ponder the wisdom of the observation *Scientia potestas est.*

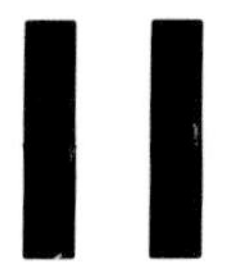

Decision-making in the Production of Power

Decision-making in the Production of Power

MILTON KATZ

It is only recently that men have begun to consider how they can reconcile human needs for energy with the finiteness of the earth. Such a reconciliation will engage all the institutions of society

The decision-makers in the production of power are many and diverse. Each has its own view of the objectives to be pursued, its own formulation of the issues, its own choice of the criteria for decision and the priorities among them, its own selection and analysis of pertinent facts and its own art of applying the criteria to the issues on the basis of the facts in pursuit of the objectives.

Who are the decision-makers? In the first instance they are the enterprises—private companies, local government agencies and such national instrumentalities as the Tennessee Valley Authority—that build and manage the power-generating facilities. The initiating enterprises draw others into the decision-making process, notably their sources of capital and commercial finance (investment and commercial banks for private companies, budget offices and appropriations committees for government agencies) and the state and national regulatory agencies that issue certificates or permits required by law or administer legal measures for safety, health or environmental protection. Often the courts, state and Federal, are drawn into the decisional process. Their authority can be invoked through statutory procedures to review the orders of regulatory agencies or through independent proceedings initiated by complainants. Still other decision-makers participate in a more general and diffuse but nonetheless critical way: Congress and the President; the legislatures and governors of the states; scientific, engineering, legal and other academic or professional organizations or groups, and the general public, which operationally tends to mean active citizen groups. At times these officials and groups are directly articulated into the decision-making process, but even when they are not, they influence the process by modifying the societal medium in which it takes place. For the purposes of the present discussion I believe it is realistic to assume a rough consensus among the various decision-makers concerning the objectives and the formulation of the main issues relating to the generation of power.

In a recent message to Congress, President Nixon described the objectives as "the blessings of both a high-energy civilization and a beautiful and healthy environment." A private power company, harassed by overstrained facilities and an angry citizenry, might describe its goals as catching up with runaway demand and gaining relief from public clamor. Although the two types of statement are not exactly congruent, they are consistent enough for practical purposes. They reflect a generally accepted projection of a need to greatly expand power-generating capacity in the U.S. during the next two decades and a heightened public sensitivity to the hazards of pollution, coupled with a growing awareness of power plants as polluters.

The issues would be identified and formulated by the respective decision-makers with a comparable consistency. The issues relate chiefly to the size, type and location of the power plants and transmission lines to be built. Concerning size, a group of experts studying electric power and the environment for the U.S. Office of Science and Technology predicted that "most of the new capacity in the next 20 years will come from some 250 huge power plants of two to three million kilowatts each." (Some 3,000 power plants are in existence today.) A power plant with a capacity of three million kilowatts would, if it burned fossil fuel, require from 900 to 1,200 acres of land, apart from the rights-of-way needed for transmission lines. A nuclear plant of comparable size would need from 200 to 400 acres, exclusive of rights-of-way. Such mammoth plants will probably need rights-of-way some 250 feet wide.

As for the type of plant, the most conspicuous issue concerns the choice between fossil-fuel and nuclear facilities, but that is only one of many issues involving difficult choices of technology, organization and method. The requirements for land, together with a need for access to low-cost fuel (in the case of a fossil-fuel plant) and water for cooling (for both kinds of plant, with nuclear facilities needing 50 percent more water than fossil-fuel plants), will engender intense competition for land with representatives of other land-using activities. Plainly the decisions concerning size and type will profoundly affect the selection and availability of sites.

The rough consensus among decision-makers ends with a bang when we pass from identifying objectives and formulating issues to choosing and applying the criteria for deciding the issues. The intensity of the bang will vary with the specificity of the decisional context. It can be muted in a large and general appraisal. The elements of division and conflict will tend to be most acute in administrative, judicial and legislative proceedings involving particular proposals. It is in the effort to sort out and apply the criteria with appropriate priorities to concrete undertakings that the implications of the criteria fully emerge.

I believe it is useful to classify the relevant criteria into two main categories. One relates to efficiency in the standard engineering and business sense, emphasizing maximum output of electric power of optimum quality at lowest cost. "Cost" is taken to mean cost to the producing enterprise, whether private or governmental. The other category relates to a comprehensive assessment of advantages and disadvantages for the society as a whole, recognizing the need for production but also emphasizing the need to minimize pollution of air, water, land and the biosphere; taking into account health, safety, recreational and aesthetic consequences and possibilities, and giving heed to the interaction of the production of electric power and other aspects of rural and industrial development.

Understandably criteria in the first category typically have dominated the calculations of the enterprises that build and manage power plants, whether private or governmental. Criteria in the second category are increasingly compelling for regulatory agencies, courts, Congress and the state legislatures, the President and the governors of states, and the general citizenry. The first category corresponds to what I shall call the enterprise outlook, the second to what I shall call the societal outlook. The terms are used in a purely descriptive sense, with no value implications intended. The enterprise outlook and the societal outlook are not mutually exclusive. On the contrary, efficient production by power companies is a matter of large social concern, and each enterprise has a practical interest—even in the strict business sense—in the health, safety and attractiveness of the surroundings of its own personnel and the communities in which they live. The elements of common interest may help to resolve the conflict, but it is the conflict itself between the two outlooks that will typify decision-making in the production of power. The conflict will center on the competing demands of power production and environmental protection.

Earlier in our history, when the prevailing value system assigned an overriding priority to the first-order effects of technology, our society would have resolved the conflict in favor of increased production almost as a matter of course. The side effects, such as pollution, would have been taken in stride. In recent years the values of the society seem to have shifted from an automatic acceptance of new technology for its own sake toward a deepening concern for environmental and other social consequences. The shift in values has begun to permeate the political process and is reflected in the President's statement of the national objectives relating to power as "both a high-energy civilization and a beautiful and healthy environment." It will remain necessary for the decision-makers to resolve the conflicts between the enterprise and the societal outlooks and the criteria that they respectively emphasize, but the old, routine and almost unconscious assumption that environmental protection must give way to production has become altogether untenable.

It does not follow that the needs of production either will or should automatically give way to environmental protection. To accommodate the competing needs decision-making in the production of power must seek an optimum balance among the competing criteria. These are brave words, but they must be translated into action. In each concrete situation, through the several efforts of the respective decision-makers and through an interaction among them, an operational meaning must be given to an "optimum balance."

The task will be intricate, arduous and difficult even in the most auspicious circumstances. In view of the multiplicity and diversity of the interests affected and the intensity and ambivalence of current feelings, the circumstances will not often be auspicious. How could the task be facilitated?

I venture to suggest a few examples of approaches, techniques and methods that might help. They involve technology assessment, the internalization of "external" or "social" costs and the use of taxation, regulatory statutes and private civil actions at law to foster technology assessment and the internalization of social costs.

Technology assessment is a procedure designed to optimize the use of technology. Modern technology, which has brought social benefits and social costs, can also multiply societal options. It can do so through the enhanced capacity to perceive and predict unintended side effects by modern analytical methods reinforced by computers and through its capacity to design many alternative means to achieve a desired objective.

Technology assessment seeks to take advantage of the variety of options to increase benefits and reduce costs. It has two main components. The first is a systematic comparative appraisal of the first-order effects of technology (electric power in the present context) in relation to the visible, discoverable and foreseeable side effects (air pollution, heating of streams or lakes and possible radioactive hazards in the present context). The second component is a search through the full range of technological possibilities for the one best designed to achieve the desired first-order effect while eliminating or minimizing the undesirable side effects.

The relation of technology assessment to protection of the environment is direct and fundamental. The environmental problem in essence is a function of (1) the growth of populations; (2) the development of technology and its application to economic and social organization; (3) the care, skill and imagination with which the potentialities of science, technology and related organization are developed and used to maximize desirable primary effects and minimize undesirable side effects, and (4) the respective priorities accorded by the society to the accomplishment of the first-order effects of technology and the elimination or reduction of the side effects. The third and fourth of these factors entail technology assessment. They are what technology assessment is all about. Technology assessment operates on certain of the primary sources of environmental problems rather than on the manifestations.

It remains to consider how technology assessment can be infused into the decisional calculations of the respective decision-makers. I begin with the power companies, along with the investment companies or banks to which they turn for capital or commercial finance. One method of fostering the incorporation of technology assessment into their decisional processes would involve the internalization of costs.

A power company contemplating a possible new plant or installation will reach its decision through the usual cost-benefit calculations. In the ordinary course of business the company will estimate the anticipated costs and benefits of a proposed installation without reference to any damage to the environment to be caused by the predictable emission of sulfur dioxide, oxides of nitrogen, particulate matter and heat or the possible leakage of radioactivity. The costs to the society entailed by such damage will be ignored by the company in its cost-accounting as a matter of course. They are treated as external or social costs. If the company's attention turns to the matter at all, it will regard the exclusion of such costs from its accounting as a routine application of standard business and accounting practice.

Generally the company will be unaware that it is able to pursue such a standard practice only because it is a beneficiary of legal doctrines that have evolved in the course of history and that can change in the continuing evolution of the society. The externality of the external costs derives neither from the fundamentals of economics nor from the nature of business. It derives from the legal system. If the legal order requires a cost arising from a company's operations to be borne by the company, the cost is internal. If the legal order requires such a cost to be borne by others, the cost is external. Damage to the environment from pollutants emitted by a power company will be an external and social cost only to the extent provided by the legal system.

I shall come later to a consideration of means whereby the legal system can internalize costs that power companies have been allowed to treat as external and therefore to exclude from their cost-benefit calculations. As we shall see, such costs can be internalized in the form of taxes, money judgments, negotiated sums payable by the companies or expenditures that the companies may be required to make by the orders of regulatory agencies. At this point I want to emphasize the practical consequences of the internalization if and when it may be effected. In particular I want to stress the implications for technology assessment and environmental protection.

The company may merely pay the internalized costs (whether they are in the form of a judgment for damages, a tax or a negotiated payment), absorbing the payment or passing it on to customers. If the company should do this and nothing more, justice might be served in a retributive or compensatory sense, but neither technology assessment nor the environment would be served at all. One such payment required by law, however, will be a portent of repeated payments to come, unless the company does something more. It can be expected that the company will seek to avoid this unhappy prospect by a change in its technology, organization or mode of operation designed to eliminate or greatly reduce the external costs by eliminating or greatly reducing the pollution. Such an improvement in technology, organization and method, if it should prove feasible, would represent a step toward the achievement of our objective: to infuse technology assessment and the optimum use of technology into the decisional processes of the power company.

The cost to the company of the improvement will replace the social cost of the pollution to the extent that the improvement is effective. The allocation of the improvement cost to the company may be definitive and complete, provisional or partial. If the improvement cost is entirely absorbed by the company, the allocation is definitive. If the improvement cost is reflected in the company's rate structure and passed on to the purchasers of electricity, the allocation to the company is merely provisional. If only a portion of the cost can be passed on to customers, the allocation to the company is partial. The allocation to the company may be partial in another sense if the company's effort to install improved technology benefits from research and development financed by a department or an agency of the government. In such a case the cost is partly borne by the department or agency.

In sum, whenever such an improvement is feasible and the cost of the improvement can be absorbed or passed on, the environment will benefit, and the production of power will also benefit or at least remain unimpaired. I believe this will be the usual case, but harder cases may arise. Let me put the hardest case in an effort to illuminate the outer reaches of the problem.

What if a company should be unable to devise any improvement that would remove or diminish the pollution? Alternatively, what if the cost of an improvement, even if it is technically feasible, should be so high and the market conditions such that the cost could neither be absorbed by the company nor passed on? To sharpen the problem let us also assume that the company operates at the highest level of managerial skill, has been constantly alert to the latest developments in technology and has maintained an energetic and imaginative program of research. In such a case the company's inability to devise an improvement or to absorb the cost could not be attributed to deficiencies in the company remediable through changes in personnel or mode of operation. What then? In the first circumstance a choice would be faced between toleration of the pollution and suppression of the company's operations. In the alternative circumstance an additional option would increase the possibilities to three: toleration of the pollution, suppression of the company's operations or a subsidy to the company out of public revenues.

Although a case so hard, confronting decision-makers with so stark a set of alternatives, will seldom arise, it is nevertheless useful as an aid to analysis. It brings to the surface the ultimate choices that to some degree are latent in most situations. It illuminates the critical importance of the procedures for decision-making, since the ultimate choices might go by default if the decisional process should be loose and confused or unbalanced. It throws light on one of the functions served by technology assessment: to make it unnecessary to face a rigorously exclusive choice between the ultimate alternatives or to mitigate the difficulty of a decision between them.

If the ultimate choices must be faced and decided, how are the decisions to be made? The nature of the decisional process must be governed by the nature of the problem. The ultimate choices are what remains to be decided after the problem has been reduced as far as practicable through scientific insight, technological skill and efficient management. The choices entail value judgments, which must be made with a sensitive regard for the deeply felt and conflicting needs and purposes of those affected. In a society that aspires to be free and open such judgments can be made only through discussion, argument, persuasion, contention, adjustment and interaction among the individuals and groups with a stake in the outcome, organized and disciplined through the democratic political process and the legal system.

The legal system can contribute powerfully to the development of suitable procedures for decision-making. (I speak of procedures in a broad sense rather than a technical one.) The legal system can also help to foster technology assessment and the internalization of costs, notably through taxation, regulatory statutes and private civil actions at law.

Taxes can be designed to foster technology assessment and the internalization of costs through incentives and pressures. They may also provide revenues to help finance research, and tax statutes may authorize accelerated write-offs for tax purposes of investments in new technology installed to reduce pollution. Excise taxes may be adopted to serve a dual purpose. Levied on the manufacture or sale of products involving a process or containing an ingredient that engenders pollution, they may stimulate the taxpayer to devise a harmless substitute process or to find a harmless substitute ingredient, while the proceeds of the tax may be allocated to the support of research for remedial measures.

The potentialities of taxes on effluents in particular have received much attention. It is contemplated that these taxes would be calculated and levied in such a way as to increase or reduce the tax burden according to the volume of pollutants emitted by the taxpayer (a power company in the present context). Some proponents of effluent taxation recommend that the tax be measured by the social cost of the pollution engendered by the taxpaying enterprises. Others advocate pragmatic standards, proposing that the tax be fixed at a level calculated to impel the taxpaying enterprise to undertake technological and managerial improvements in an effort to reduce the tax by diminishing the pollution. The design and the administration of such taxes could encompass many variations. Whatever the precise arrangement, the intended primary functions would be to internalize costs and promote technology assessment.

Under the typical statutes regulating power companies supervision is exercised by regulatory commissions in rate-setting, investment and related accounting practices. The criteria of technology assessment and cost internalization could readily be incorporated into the decisional processes of a regulatory commission. Authority to do so could be derived from the governing statute through appropriate interpretations or provided by legislative amendments. In deciding whether to grant or withhold a certificate authorizing new construction the commission could condition its approval on a showing that the proposed new plant or equipment was designed to minimize pollution while achieving efficient production. (An approximation of such a requirement is provided in an Administration bill pending in Congress that would establish new procedures for selecting sites for new power plants. The bill would authorize state, regional and Federal certifying bodies to issue certifications of site and facility for bulk power-supply facilities "if such bodies find, after having considered available alternatives, that the use of the site or route will not unduly impair environmental values and will be reasonably necessary to meet electric power needs.")

To encourage a company to seek environmental protection along with efficient production through improved technology, organization and methods of operation, the commission could indicate its willingness to reflect the improvement costs in the company's rate schedule. The commission could broaden the scope of its inquiries and obtain increased assistance for its own understanding of the issues and their implications by inviting individuals and groups representing diverse interests and outlooks to take part in its proceedings. It might adopt such measures on its own initiative, or it might be prodded into taking them by judicial review or legislative intervention.

Statutes designed to protect the environment against injury from any source can also promote technology assessment in the production of power. Recent legislation has multiplied environmental-protection statutes that fix—or authorize designated agencies to fix—standards for emissions, air quality or water quality. Power companies usually come within their reach, although the statutes are intended generally to curb pollution rather than specifically to regulate power companies.

As applied to a power company, such a statute would require the company to adjust its operations to fit limits contained within or derived from the statutory standards. The adjustment would involve choices comparable to those entailed by the internalization of costs through a tax or a money judgment, with one exception. In the first instance the company would be denied the alternative of merely paying the tax or the judgment obligation. The company's alternatives would be to curb emissions through improvements in technology, organization or mode of operation, absorbing the improvement cost or passing it on; to curtail its production; to obtain a relaxation of the statutory standard, or to violate the statute and suffer the statutory penalty. If the penalty should be a fine (as is typically the case), the company could be said to have a final alternative of simply making a monetary payment.

Private civil actions at law have a special significance in that they provide an outlet for efforts by independent citizens. Such actions offer a means whereby the multiple initiatives of private citizens, individually or in groups, can be brought to bear on technology assessment, the internalization of costs and environmental protection. They constitute a channel through which the diverse interests, outlooks and moods of the general public can be given expression.

The current popular concern over the environment has stimulated private civil actions of two main types. In one type the action is aimed directly at an enterprise that causes pollution. In the other type the action tries to reach offending enterprises indirectly through a regulatory commission that supervises their operations, seeking to hold the commission to its duties, to curb improper action on its part and to galvanize it to appropriate action. Private actions of the first type fall predominantly within the range of tort law, those of the second type within the sphere of administrative law.

Although tort actions based on negligence have a limited use, the potentialities of tort law for technology assessment are found mainly in the doctrines of nuisance, liability for "abnormally dangerous" or "extrahazardous" activities and strict products liability. For power companies it is chiefly the doctrine of nuisance and secondarily the doctrine of liability for injuries caused by extrahazardous activities that are likely to prove useful. These doctrines provide remedies for injuries that arise from the very nature of the defendant's operations, without regard to its intention (as intention is ordinarily understood in tort law) or any negligence on its part. The decisions of courts in such cases are reached and stated in technical terms. When the judicial opinions are read with an eye informed by an understanding of technology assessment, a remarkable degree of confluence both in mode of thought and in practical consequences can be discerned between the tort doctrines and technology assessment.

A successful tort action leads to a judgment for damages or an injunction against the defendant company. The options for the company arising from a judgment against it for damages have already been explained. An injunction typically will prohibit the defendant from emitting pollutants or from continuing such of its operations as cause pollution. (An injunction of the latter type would be highly unlikely in the case of a power company.) Confronted by such an injunction, the defendant must either devise a suitable alternative technology, organization or method, curtail its production or negotiate with the plaintiff for an agreement to dissolve the injunction in return for a stipulated payment. The implications of such choices have been explored earlier in this article. In sum, the significance of a tort judgment for technology assessment in the production of power will be determined by the extent to which it spurs the defendant power company to improve its technology and management.

In the existing state of tort law and administrative law there are obstacles that limit the scope of private civil actions to protect the environment against polluting enterprises or laggard regulatory agencies. The obstacles can be reduced and the constructive potentialities of such actions can be enhanced by incremental judicial improvement and by remedial and supplementary legislation.

The Authors
Bibliographies
Index

The Authors

CHAUNCEY STARR is dean of the School of Engineering and Applied Science at the University of California at Los Angeles. Before he assumed that position in 1967 he had spent 20 years in industry, including service as vice-president of North American Rockwell and president of its Atomics International Division. Starr received his bachelor's degree in electrical engineering from Rensselaer Polytechnic Institute in 1932 and his Ph.D. in physics from the same institution in 1935. He did work in the field of high pressure at Harvard University and in cryogenics at the Massachusetts Institute of Technology, was with the Radiation Laboratory of the University of California at Berkeley during World War II as part of the Manhattan District and worked after the war on the development of nuclear propulsion for rockets and ramjets, on miniaturizing nuclear reactors for space missions and on developing nuclear plants for the production of electric power. Among other activities he is vice-president of the National Academy of Engineering and a member of the President's Task Force on Science Policy.

FREEMAN J. DYSON is professor in the School of Natural Sciences of the Institute for Advanced Study. Born in England, he was educated at the University of Cambridge and did operational research for the Royal Air Force in World War II. He joined the Institute for Advanced Study in 1953 after two years as professor of physics at Cornell University.

M. KING HUBBERT is research geophysicist with the U.S. Geological Survey. He writes: "I was born in central Texas only 30 years after the cessation of Comanche Indian raids on the local frontier communities. I attended one- and two-teacher country schools until about the ninth grade and then went for a year to a small private school, from which I was graduated in 1921. I spent the next two years at Weatherford College, a small junior college in Texas, before leaving in May, 1923, to work in the wheat fields of Oklahoma and Kansas and as a gandy dancer on the Union Pacific Railroad, en route to the University of Chicago. In Chicago I was successively a telephone installer, a postal clerk, a waiter, a checkroom attendant, a supernumerary in *Romeo and Juliet* and in various operas and a camp cook in Wisconsin in addition to attending the university." Hubbert took his bachelor's, master's and doctor's degrees in geology and physics at Chicago and taught geology and geophysics at Columbia University for 10 years. He was with the Shell Oil Company for 20 years before he joined the Geological Survey in 1964.

DAVID M. GATES is professor of botany at the University of Michigan and director of the university's Biological Station. From 1965 until this year he was director of the Missouri Botanical Gardens and professor of biology at Washington University in St. Louis, Gates began his career as a physicist. Having received his bachelor's, master's and doctor's degrees in physics from the University of Michigan, he spent the first decade of his professional career in research on atmospheric physics. During the second decade he continued work in physics while taking an interest in ecology and plant physiology. In the third decade he has assumed responsibilities that are primarily within biology.

WILLIAM B. KEMP is lecturer in geography at McGill University. "Since 1962," he writes, "much of my time has been spent doing field research among the eastern Eskimos of the Canadian Arctic. I enjoy long periods of field research, but of greater importance is the need to collect detailed information on a way of life that is rapidly disappearing from the North American scene." Kemp was graduated from Miami University in Ohio in 1959 and developed his interest in the Arctic which doing graduate work at Michigan State University. He received his master's degree there in 1961 and expects to receive his Ph.D. "when I finish analyzing the information collected from 1967 to 1970." Before going to McGill in 1970 he taught at the State University of New York at Binghamton.

ROY A. RAPPAPORT is associate professor of anthropology at the University of Michigan. He did not take up anthropology professionally until 10 years after being graduated from Cornell University with a degree in hotel administration. During most of those years he owned an inn in Lenox, Mass. "In 1959," he writes, "I left the resort business to enter graduate studies in anthropology at Columbia University, receiving a Ph.D. from that institution in 1966 after doing archaeological fieldwork in the Society Islands and ethnographic fieldwork in New Guinea." His main interests are the ecology of nonindustrial people and religion, and particularly the relation between ecology and religion, which he examined in the book *Pigs for the Ancestors: Ritual in the Ecology of a New Guinea People.*

EARL COOK is professor of geology and geography at Texas A&M University and also associate dean of the College of Geosciences and director of the Environmental Quality Program at the university. He was graduated from the University of Washington in 1943 with a degree in mining engineering. After service in World War II he returned to the university, receiving his master's degree in geology in 1947 and his Ph.D. in geology in 1954. From 1951 to 1964 he was at the University of Idaho, starting as assistant professor and ending as dean of the College of Mines. He then spent almost three years as executive secretary of the Division of Earth Sciences of the National Academy of Sciences–National Research Council before going to Texas. He writes of his interest in "the evolution of concepts, particularly in Western thought, of man's relation to his environment and how these concepts are related to man's impact on the environment."

CLAUDE M. SUMMERS is consulting professor of electric-power engineering at Rensselaer Polytechnic Institute. After receiving degrees from the University of Colorado he was with the General Electric Company from 1927 to 1959, including service as manager of the laboratory at the Fort Wayne works. He then spent nine years at Oklahoma State University as professor of electrical engineering. On retiring in 1968 he went to Rensselaer to assist in the development of a new electrical-machinery laboratory.

DANIEL B. LUTEN is lecturer in geography at the University of California at Berkeley. Before going to Berkeley in 1961 he was with the Shell Oil Company for 25 years, working at what he calls "a wholesome variety of tasks." From 1948 to 1950 he was on leave to serve as technical adviser to the chief of the natural resources section in the civil administration of occupied Japan. Luten was graduated from Dartmouth College in 1929 and obtained his Ph.D. in chemistry from Berkeley in 1933. He writes: "Have progressed from knowing perhaps everything about what (for example petroleum cresylic acids) my friends assured me was nothing to maybe not knowing nothing about everything, but at least nothing about everything I've tried to look into carefully."

MYRON TRIBUS and EDWARD C. McIRVINE are with the Xerox Corporation; Tribus is senior vice-president in the Business Products Group and general manager of research and engineering, and McIrvine is manager of the physics research laboratory. Before joining Xerox last year Tribus was with the U.S. Department of Commerce as assistant secretary for science and technology. From 1961 to 1969 he was dean of the School of Engineering at Dartmouth College, having gone there following eight years as professor of engineering at the University of California at Los Angeles, where he received his Ph.D. in engineering in 1949. In 1958 he was host, moderator and writer for the Columbia Broadcasting System's television show "Threshold." McIrvine, who was graduated from the University of Minnesota in 1954 and took his Ph.D. in theoretical physics at Cornell University in 1959, joined Xerox in 1969 after nine years with the Ford Motor Company.

MILTON KATZ is Henry L. Stimson Professor of Law at Harvard University and director of Harvard's International Legal Studies Program. He was graduated from Harvard College in 1927, spent a year crossing central Africa on an anthropological expedition for the Peabody Museum of Harvard and then entered the Harvard Law School, receiving his LL.B. in 1931. After several years of service with the U.S. Government he joined the Harvard Law School faculty in 1939. Recalled to the Government in 1941, he returned to Harvard in 1946 but in 1948 was appointed general counsel for the Economic Cooperation Administration in Europe. In 1950–1951 he became head of the Marshall Plan in Europe, with the rank of ambassador. From 1951 to 1954 he was vice-president of the Ford Foundation, returning to Harvard in 1954. Among many other activities he is chairman of the board of trustees of the Carnegie Endowment for International Peace.

Bibliographies

Readers interested in further reading on the subjects covered by articles in this issue may find the lists below helpful.

ENERGY AND POWER

ENERGY IN THE FUTURE. Palmer Cosslett Putnam. D. Van Nostrand Company, Inc., 1953.

ENERGY IN THE AMERICAN ECONOMY, 1850–1975: AN ECONOMIC STUDY OF ITS HISTORY AND PROSPECTS. Sam H. Schurr and Bruce C. Netschert. The Johns Hopkins Press, 1960.

RESOURCES IN AMERICA'S FUTURE: PATTERNS OF REQUIREMENTS AND AVAILABILITIES 1960–2000. Hans H. Landsberg, Leonard L. Fischman and Joseph L. Fisher. The Johns Hopkins Press, 1963.

DIRECT USE OF THE SUN'S ENERGY. Farrington Daniels. Yale University Press, 1964.

THE WORLD ELECTRIC POWER INDUSTRY. N. B. Guyol. University of California Press, 1969.

CONTROLLED NUCLEAR FUSION: STATUS AND OUTLOOK. David J. Rose in *Science*, Vol. 172, No. 3985, pages 797–808; May 21, 1971.

ENERGY IN THE UNIVERSE

GRAVITATION THEORY AND GRAVITATIONAL COLLAPSE. B. Kent Harrison, Kip S. Thorne, Masami Wakano and John Archibald Wheeler. The University of Chicago Press, 1965.

THE DYNAMICS OF DISK-SHAPED GALAXIES. C. C. Lin in *Annual Review of Astronomy and Astrophysics,* Vol. 5, pages 453–464; 1967.

GRAVITATION AND THE UNIVERSE: JAYNE MEMORIAL LECTURE FOR 1969. Robert H. Dicke. American Philosophical Society, 1970.

THE ENERGY RESOURCES OF THE EARTH

MAN AND ENERGY. A. R. Ubbelohde. Hutchinson's Scientific and Technical Publications, 1954.

ENERGY FOR MAN: WINDMILLS TO NUCLEAR POWER. Hans Thirring. Indiana University Press, 1958.

ENERGY RESOURCES. M. King Hubbert. National Academy of Sciences–National Research Council, Publication 1000-D, 1962.

RESOURCES AND MAN: A STUDY AND RECOMMENDATIONS. Committee on Resources and Man. W. H. Freeman and Company, 1969.

ENVIRONMENT: RESOURCES, POLLUTION AND SOCIETY. Edited by William W. Murdoch. Sinauer Associates, 1971.

THE FLOW OF ENERGY IN THE BIOSPHERE

FUNDAMENTALS OF ECOLOGY. Eugene P. Odum. W. B. Saunders Company, 1959.

ENERGY EXCHANGE IN THE BIOSPHERE. David M. Gates. Harper & Row, Publishers, 1962.

PHYSICAL CLIMATOLOGY. William D. Sellers. The University of Chicago Press, 1965.

ENERGY FLOW IN BIOLOGY. Harold J. Morowitz. Academic Press, 1968.

CONCEPTS OF ECOLOGY. Edward Kormondy. Prentice-Hall, Inc., 1969.

THE FLOW OF ENERGY IN A HUNTING SOCIETY

THE NETSILIK ESKIMOS: SOCIAL LIFE AND SPIRITUAL CULTURE. Knud Rasmussen in *Report of the Fifth Thule Expedition, 1921–24, No. 8.* Copenhagen: Gyldendalake Boghandel, 1931.

THE ESKIMOS: THEIR ENVIRONMENT AND FOLKWAYS. Edward Moffat Weyer, Jr. Yale University Press, 1932.

MAN THE HUNTER. Edited by Richard B. Lee and Irven DeVore. Aldine Publishing Company, 1969.

THE FLOW OF ENERGY IN AN AGRICULTURAL SOCIETY

ECOLOGICAL ENERGETICS. John Phillipson. Edward Arnold Publishers, 1966.

PERSPECTIVES IN ECOLOGICAL THEORY. Ramón Margalef. The University of Chicago Press, 1968.

PIGS FOR THE ANCESTORS: RITUAL IN THE ECOLOGY OF A NEW GUINEA PEOPLE. Roy A. Rappaport. Yale University Press, 1968.

ENVIRONMENT, POWER, AND SOCIETY. Howard T. Odum. Wiley-Interscience, 1971.

THE FLOW OF ENERGY IN AN INDUSTRIAL SOCIETY

ENERGY IN THE UNITED STATES: SOURCES, USES, AND POLICY ISSUES. Hans H. Landsberg and Sam H. Schurr. Random House, 1968.

AN ENERGY MODEL FOR THE UNITED STATES, FEATURING ENERGY BALANCES FOR THE YEARS 1947 TO 1965 AND PROJECTIONS AND FORECASTS TO THE YEARS 1980 AND 2000. Warren E. Morrison and Charles L. Readling. U.S. Department of the Interior, Bureau of Mines, No. 8384, 1968.

THE ECONOMY, ENERGY, AND THE ENVIRONMENT: A BACKGROUND STUDY PREPARED FOR THE USE OF THE JOINT ECONOMIC COMMITTEE, CONGRESS OF THE UNITED STATES. Environmental Policy Division, Legislative Reference Service, Library of Congress. U.S. Government Printing Office, 1970.

ENERGY CONSUMPTION AND GROSS NA-

TIONAL PRODUCT IN THE UNITED STATES: AN EXAMINATION OF A RECENT CHANGE IN THE RELATIONSHIP. National Economic Research Associates, Inc., 1971.

THE CONVERSION OF ENERGY

EFFICIENCY OF THERMOELECTRIC DEVICES. Eric T. B. Gross in *American Journal of Physics,* Vol. 29, No. 1, pages 729–731; November, 1961.

ELECTRICAL ENERGY BY DIRECT CONVERSION. Claude M. Summers. Publication No. 147, The Office of Engineering Research, Oklahoma State University, March, 1966.

APPROACHES TO NONCONVENTIONAL ENERGY CONVERSION EDUCATION. Eric T. B. Gross in *IEEE Transactions on Education,* Vol. E-10, No. 2, pages 98–99; June, 1967.

THE ECONOMIC GEOGRAPHY OF ENERGY

A HISTORY OF TECHNOLOGY. Edited by Charles Singer, E. J. Holmyard and A. R. Hall. Oxford University Press, 1954–1958.

NATIONAL POWER SURVEY: GUIDELINES FOR GROWTH OF ELECTRIC POWER INDUSTRY. Federal Power Commission, U.S. Government Printing Office, 1964.

INTERNATIONAL PETROLEUM ENCYCLOPEDIA. The Petroleum Publishing Co., 1968.

ENERGY AND INFORMATION

INFORMATION THEORY AS THE BASIS FOR THERMOSTATICS AND THERMODYNAMICS. Myron Tribus in *Transactions of the American Society of Mechanical Engineers, Series E, Journal of Applied Mechanics,* Vol. 83, pages 1–8; March, 1961.

SCIENCE AND INFORMATION THEORY. Leon Brillouin. Academic Press Inc., 1962.

THE MATHEMATICAL THEORY OF COMMUNICATION. Claude E. Shannon and Warren Weaver. University of Illinois Press, 1963.

WHY THERMODYNAMICS IS A LOGICAL CONSEQUENCE OF INFORMATION THEORY. Myron Tribus, Paul T. Shannon and Robert B. Evans in *A.I.Ch.E. Journal,* Vol. 12, No. 2, pages 244–248; March, 1966.

A PROOF THAT ESSERGY IS THE ONLY CONSISTENT MEASURE OF POTENTIAL WORK (FOR WORK SYSTEMS). Robert B. Evans. Ph.D. Thesis, Dartmouth College, 1969. University Microfilms, Ann Arbor, Mich.

DECISION-MAKING IN THE PRODUCTION OF POWER

TECHNOLOGY: PROCESSES OF ASSESSMENT AND CHOICE, A REPORT OF THE NATIONAL ACADEMY OF SCIENCES. House Committee on Science and Astronautics. U.S. Government Printing Office, July, 1969.

THE FUNCTION OF TORT LIABILITY IN TECHNOLOGY ASSESSMENT. Milton Katz. Harvard University Program on Technology and Society, Reprint No. 9, 1970.

WORK GROUP ON ENERGY PRODUCTS. *Man's Impact on the Global Environment: Assessment and Recommendations for Action.* The M.I.T. Press, 1970.

WORK GROUP ON IMPLICATIONS OF CHANGE. *Man's Impact on the Global Environment: Assessment and Recommendations for Action.* The M.I.T. Press, 1970.

ELECTRIC POWER AND THE ENVIRONMENT. The Energy Policy Staff, Office of Science and Technology. U.S. Government Printing Office, August, 1970.

Index